28 Advances in Biochemical Engineering/ Biotechnology

Managing Editor: A. Fiechter

Microbial Activities

With Contributions by
I. N. Gogotov, E. N. Kondratieva,
J. H. Luong, F. Parisi,
B. Sonnleitner, B. Volesky

With 46 Figures and 46 Tables

Springer-Verlag
Berlin Heidelberg New York Tokyo
1983

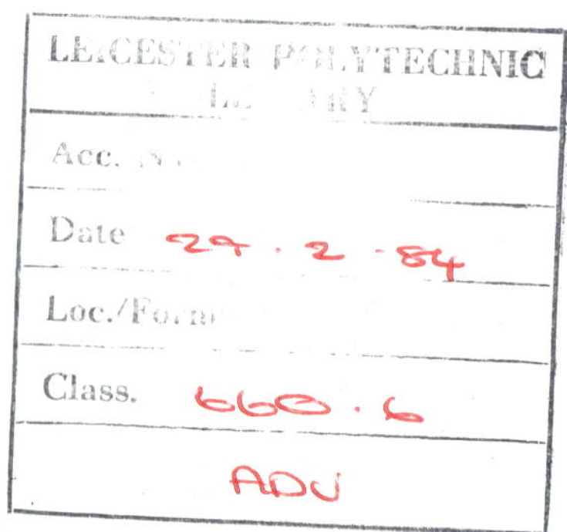

ISBN 3-540-12791-7 Springer-Verlag Berlin Heidelberg New York Tokyo
ISBN 0-387-12791-7 Springer-Verlag New York Heidelberg Berlin Tokyo

Library of Congress Catalog Card Number 72-152360
Printed in GDR

2152/3020-543210

Table of Contents

Heat Evolution During the Microbial Process — Estimation, Measurement, and Applications

J. H. T. Luong
Central Research and Development Department John Labatt Ltd. London, Ontario, N6A 4M3, Canada

B. Volesky
Department of Chemical Engineering, McGill University, Montreal, Quebec, H3A 2A7, Canada

Estimation and measurement techniques of microbial heat evolution during a microbial process are concisely discussed in this review.

In a well-defined medium and when the products are specified, the heat evolution quantity can be estimated directly from the heats of combustion of organic substrate, products, and biomass. Theoretical estimations of the heats of combustion of organic substrates or products can be very accurate. The heat of combustion of microbial cells is usually determined by burning microbial cells in an oxygen bomb calorimeter. However, when the composition of the biomass is known, the heat of combustion of the cell can be calculated by one of the following procedures: the Giese method, the Dulong equation, and the method of Mennett and Nakayama.

In a complex medium and when either non-cellular or cellular products are not completely specified, heat evolved during the microbial activity can be directly measured by microcalorimetry, dynamic calorimetry, and continuous calorimetry, respectively. Microcalorimetry usually involves the use of an adiabatic calorimeter, a thermal fluxmeter or a flow calorimeter.

On a bench-scale, dynamic calorimetry of continuous calorimetry can be applied to monitor the quantity of heat evolution during the fermentation. While the continuous calorimetric technique is used to continuously measure the heat evolution, dynamic calorimetry only provides intermittent data and requires continuous attention during the course of the process.

Based upon the heat evolution parameter; the biomass concentration, the oxygen consumption, the carbon dioxide production, and the substrate utilization, the energy requirements for growth, product formation and maintenance purposes are correlated to the biomass energetic yield and the product energetic yield. Under aerobic conditions, the fraction of the substrate energy evolved as heat is equal to the fraction of available electrons transferred to oxygen. As a consequence, either the heat evolution data or the oxygen consumption data is quite useful for monitoring a microbial process.

The heat evolution data can be used as a kind of a "heat probe" employed in process control and in on-line optimization of either aerobic or anaerobic systems. The heat evolution should be used together with the other microbial process parameters for estimating the cellular metabolic activity and bioreaction kinetics and for checking the consistency of other experimental data collected during the experiment.

1 Introduction

The growth of microorganisms is accompanied by the production of heat, regardless of whether the system is aerobic or anaerobic or whether the final product is biomass or metabolites. A large part of the heat is generated during the degradation of the organic substrate, which serves as a carbon and energy source. The catabolic process is always associated with a significant free-energy decrease. Part of the released energy is conserved in high-energy bonds of adenosine triphosphate (ATP) or other energy-storage compounds, providing energy for biosynthesis of cellular components and other microbial activities as required. The rest of the substrate's original energy is released as heat. In addition to the heat production during the catabolism of the energy-rich nutrient molecules, most of the ATP energy is also liberated as heat during its utilization in the cellular activities, providing support for the microbial growth and other cell functions.

The amount of heat produced is dependent upon the type of catabolic pathway through which the organic substrate is metabolized. It is also dependent upon the energetic coupling of energy-storage compound (ATP, etc.) generation and cell biosynthesis. Variations in the microbial heat parameter reflect the type of metabolic activities of the cell, as well as the degree of perfection with which the cellular metabolism and anabolism are completed. This indicates that the heat generation during the activity of microbial culture may be used to evaluate the regulatory mechanism of the cellular energy metabolism. It may also be very helpful in estimating the energetic efficiency of catabolic pathways, the heat energetic yield and the efficiency of energy recovery by the cells.

However, the metabolic heat is still an under-utilized parameter for assessing the degree of microbial activities. This could be explained by the fact that the heat evolution is not usually monitored due to the difficulties and/or complexities involved in the measurement of such an elusive parameter. Also, there is a lack of good interpretation of the heat evolution quantity. In most cases, the rate of metabolic heat release has been correlated with the rate of oxygen consumption by the growing culture [1,2,3,4]. This type of correlation is not applicable to anaerobic microbial systems and it only represents a narrow aspect of cellular metabolic activity.

A successful attempt was reported by Belaich et al. [5] to correlate the quantity of heat released with the quantities of substrate utilization and product (ethanol) biosynthesis. The metabolic heat parameter has also been used to predict the maximum value of the specific rate of substrate utilization and the Michaelis-Menten constant of the ethanol biosynthesis system.

Many investigators have attempted the correlation between the heat evolution and cellular metabolic activity as a means of gaining insight into the thermodynamics and energetics of growth [6,7]. However, little attention has been given to the possibility of measuring heat production for routine assessment of microbial kinetics. Mou and Cooney [8], applying the dynamic calorimetric technique to an antibiotic biosynthesis in complex media, found thermal measurements useful for monitoring cell growth and as a physiological parameter throughout the bioconversion process.

Recently, Wang et al. [9], Luong and Volesky [10], Volesky et al. [11] have used the parameter of heat produced to indirectly assess the concentration and rate of growth of cells during a microbial process.

Heat and mass balances have been presented together with three regularities which Minkevich and Eroshin [12,13,14,15] have identified and quantified. Erickson [16,17], and Erickson and co-workers [18,19,20,21] have applied the heat and mass balance regularities to specific culture conditions such as continuous cultivations, batch biosystems, and processes with extracellular products. The above-mentioned investigators have demonstrated that material and energy balance regularities can be used to analyze the process energetics.

Applications of material and energy balances and associated regularities in on-line data analysis have been attempted [22]. This approach is particularly important in on-line data analysis where only limited microbial process data are available.

In view of this, the main purpose of this paper is to illustrate some important applications of the heat evolution quantity as a microbial process parameter. In aerobic systems, it can be used to predict the cellular metabolic activity, the biomass concentration, the oxygen uptake rate, and the organic substrate utilization rate. The heat evolution parameter can also be used, together with either oxygen uptake data or carbon dioxide respiration data, for checking the consistency of experimental data collected during the experiment.

In anaerobic systems, the applicability of the heat evolution data as an analytical tool for the routine assessment of microbial kinetics will be discussed. Theoretical estimation and experimental measurements of the heat evolution data will also be presented in detail.

2 Theoretical Calculation of Microbial Metabolic Heat Evolution

The microbial heat evolution, ΔQ, from an exothermic microbial process is calculated as follows:

$$\Delta Q = (-\Delta H_s)(-\Delta S) + (-\Delta H_n)(-\Delta N) - (-\Delta H_c)(\Delta X) - \Sigma(-\Delta H_{pi})(\Delta P_i) \quad (1)$$

where

$-\Delta H_s$ = heat of combustion of carbon substrate
$-\Delta H_n$ = heat of combustion of nitrogen source
$-\Delta H_c$ = heat of combustion of microbial cells
$-\Delta H_{pi}$ = heat of combustion of ith product
$-\Delta S$ = amount of substrate utilized
$-\Delta N$ = amount of nitrogen utilized
ΔX = amount of microbial cells produced
ΔP_i = amount of ith product produced

The combustion heat values of various substances, $(-\Delta H_s)$ and $(-\Delta H_{pi})$ can be obtained from many different sources. In the absence of experimental data, the heat of combustion of an organic substance can be calculated by the following procedures [23]:

— The electrons in the C—C bond or C—H bond of a respective molecule produce heat energy of 108.99 kJ per equivalent of electrons.

— The electrons in C = 0, CHOH, and CH_2 OH in the respective molecule produce additional heat energy of 81.59, 54.39, and 54.39 kJ per equivalent of electrons respectively.

Generally, the values predicted on the basis of the above assumption appear to compare well with the experimental values (Table 1). The heat of combustion of microbial cells $(-\Delta H_c)$ is usually unknown. It is experimentally determined by burning

Table 1. The values of heat of combustion of some chemicals theoretical vs experimental

Chemical	Heat of combustion (kJ mol^{-1})		% Difference from the experimental value
	Measurement [24]	Predicted	
Methane	890.3		— 2.07
Methanol	726.8	708.3 (108.99 × 6 + 54.39)	— 2.53
Ethanol	1366.1	1362.3 (108.99 × 12 + 54.39)	— 2.76×10^{-1}
Glycerol	1664.4	1689.1 (108.99 × 14 + 54.9 × 3)	1.48
Formaldehyde	561.1	517.6 (108.99 × 4 + 81.59)	— 7.75
Acetaldehyde	1166.5	1171.5 (108.99 × 10 + 81.59)	4.3×10^{-1}
Acetone	1825.5	1829.7 (108.99 × 16 + 81.59)	2.29×10^{-1}
Formic acid	263.2	218 (108.99 × 2)	—17.17
Acetic acid	872.8	871.9 (108.99 × 8)	— 9.6×10^{-2}
Lactic acid	1364	1362.3 (108.99 × 12 + 54.39)	— 1.13×10^{-1}
Pyruvic acid	1171.5	1171.5 (108.99 × 10 + 81.59)	0
Tartaric acid	1151	1198.7 (108.99 × 10 + 54.39 × 2)	4.14
Maleic acid	1338.9	1362.3 (108.99 × 12 + 54.39)	1.72
Succinic acid	1494.1	1525.9 (108.99 × 14)	2.13
Fumaric acid	1338.9	1305.4 (108.99 × 12)	— 2.5
Xylose	2349.3	2355.6 (108.99 × 20)	0.27
Galactose	2806.2	2887.8 (108.99 × 24 + 54.39 × 5)	2.91
Glucose	2815.8	2887.8 (108.99 × 24 + 54.39 × 5)	2.56
Rhamnose	3005.4	3050.1 (108.99 × 26 + 54.39 × 4)	1.49
Maltose	5649.2	5666.8 (108.99 × 48 + 54.39 × 8)	3.11×10^{-1}

microbial cells in an oxygen bomb calorimeter. According to Prochazka et al. [25], the mean calorific content of the various microorganisms they tested was around 22.6 kJ per g cells on an ash-free, dry weight basis; the range of variation was from 20.9 to 25.1 kJ per g cells.

When the composition of the cells is specified, the heat of combustion of the cells can be calculated by the following methods:

1. A revised version of the Dulong equation [26] can be used to calculate the calorific content of microorganisms.

$$\frac{\text{kJ}}{\text{g cells}} = 33.76\,\text{C} + 144.05\left[\text{H} - \frac{0}{8}\right] \tag{2}$$

Where C, H, and O are the weight fractions of carbon, hydrogen and oxygen of the microbial cells.

2. A similar equation used by Giese [27] to calculate the heat of combustion of an organic compound can be applied to calculate the calorific content of microorganisms.

$$\text{kJ g}^{-1} = \frac{460.24 \times (\text{RL}) \times N_c}{\text{mol. wt. of cells}} \tag{3}$$

Where the reduction level of the cells (RL) is expressed by the following formula:

$$\text{RL} = \frac{2N_c + (N_H/2) - N_O}{2N_c} \tag{4}$$

N_c, N_H, and N_0 are the numbers of carbon, hydrogen, and oxygen atoms in an empirical molecular formula for cells.
The Giese equation appears to be more applicable than the Dulong equation for estimating the calorific content of microbial cells [26].

3. An interesting method uses also developed by Mennett and Nakayama [28] to estimate the calorific content of microorganisms. This method is based upon the sum of the heats of combustion of the macromolecular components of the cells: 22.59 kJ per g protein 38.9 kJ per g lipid, 17.57 kJ per g carbohydrate and 14.64 kJ per g nucleic acids. This theoretical analysis estimated a value of 22.17 kJ per g cells of ash-free cells, dry weight for *Pseudomonas fluorescens* which agreed very well with the experimental value obtained by Prochazka et al. [25].

3 Estimation of the Microbial Metabolic Heat Evolution Based on Respiration

In the cases of aerobic cultivation without producing noncellular product, i.e. $\Sigma\Delta P_i = 0$, the following equation can be rewritten based upon Eq. (1):

$$\Delta Q = (-\Delta H_s)(-\Delta S) + (-\Delta H_n)(-\Delta N) - (-\Delta H_c)(\Delta X) \tag{5}$$

Minkevich and Eroshin [2] found that the heat of combustion of an organic substance and dried could be calculated by multiplying the proportionality constant of 451.8 kJ per mol O_2 by the amount of oxygen required for the oxidation of each substance.

$$(-\Delta H_s) = (-\Delta H_0)\, A_s \tag{6}$$

$$(-\Delta H_c) = (-\Delta H_0)\, A_c \tag{7}$$

where

$$(-\Delta H_0) = 451.8 \text{ kJ per mol } O_2 \tag{8}$$

If it is further assumed that the heat of combustion of nitrogeneous substance could also be estimated similarly:

$$(-\Delta H_n) = A_N(-\Delta H_0) \tag{9}$$

from Equations (5)–(9), it follows that:

$$\Delta Q = [A_s(-\Delta S) + A_N(-\Delta N) - A_c(\Delta X)]\,(-\Delta H_0) \tag{10}$$

obviously,

$$(-\Delta O_2) = A_s(-\Delta S) + A_N(-\Delta N) - A_c(\Delta X) \tag{11}$$

where $(-\Delta O_2)$ = Quantity of oxygen required for aerobic respiration

Therefore,

$$\Delta Q = (-\Delta H_0)\,(-\Delta O_2) \tag{12}$$

Cooney et al. [1], using the *in situ* dynamic calorimetric technique directly measuring the metabolic heat and the oxygen uptake rate, found the following relationship:

$$\Delta Q = (518.8 \text{ kJ per mol } O_2)\,(-\Delta O_2) \tag{13}$$

Luong and Volesky [4], developing the continuous calorimetric technique of directly measuring the heat evolution of *A. niger*, *E. coli*, *C. lipolytica*, *C. intermedia*, and *C. utilis*, found the following correlation:

$$\Delta Q = (460.2 \text{ kJ per mol } O_2)\,(-\Delta O_2) \tag{14}$$

Based on the concept of Mayberry et al. [29] and Payne [30] who advocated the use of heat of combustion per available electron equivalent, Imanaka and Aiba [3] developed the following relationship:

$$\Delta Q = (-\Delta H_0^*)\,(-\Delta O_2) \tag{15}$$

where the value of $(-\Delta H_0^*)$ may be taken as 111 (kJ) per av.e) X 4 (av.e per mol O_2) = 444 kJ per mol O_2.

The rate of oxygen consumption is calculated from the mass balances of respective gas between inlet and outlet air streams of a bioreactor [31]:

$$\dot{q}_{O_2} = \frac{F_N}{V_L}\left[\frac{P_{O_2}^{in}}{P_T - P_{O_2}^{in} - P_{CO_2}^{in} - P_w^{in}} - \frac{P_{O_2}^{out}}{P_T - P_{O_2}^{out} - P_{CO_2}^{out} - P_w^{out}}\right] \quad (16)$$

where

P_T = total pressure
$P_{O_2}^{in}$, $P_{O_2}^{out}$ = partial pressure of oxygen in inlet and outlet gas
$P_{CO_2}^{in}$, $P_{CO_2}^{out}$ = partial pressure of carbon dioxide in inlet and outlet gas
P_w^{in}, P_w^{out} = partial pressure of water in inlet and outlet gas
F_N = molal flow rate of inert gas (N_2)
V_L = volume of liquid broth

The total oxygen consumed during a microbial process is then calculated by an integrating method:

$$(-\Delta O_2) = \int_0^t \dot{q}_{O_2}\, dt \quad (17)$$

In the absence of experimental data and when the elemental compositions of microbial cells are specified, the amount of oxygen required per unit weight of microbial cells produced can be estimated by the following formula [32]:

$$\frac{g\ O_2}{g\ cell} = 16\left[\frac{2C + (H/2) - O}{Y_{x/s.(Mol.\ wt.)}} + \frac{0'}{1600} - \frac{C'}{600} + \frac{N'}{933} - \frac{H'}{200}\right] \quad (18)$$

where

— C, H, and O represent the number of atoms of carbon, hydrogen, and oxygen, respectively, in each molecule of carbon source.
— C′, H′, O′, and N′ represent the percentage of carbon, hydrogen, oxygen, and nitrogen, respectively in the cells.
— Mol. wt. represents the molecular weight of the carbon source.
— $Y_{x/s} = \Delta X/-\Delta S$, represents the yield of cells based on carbon source.

Equation (18) is only valid when the nitrogen source is ammonia and the only products of metabolisms are the cells, carbon dioxide, and water. Growth yield based upon oxygen, $Y_{x/0} = \Delta X/(-\Delta O_2)$, gram cell produced per gram of oxygen consumed of various microorganisms growing aerobically in minimal media is summarized in Table 2.

4 Direct Measurement of the Microbial Metabolic Heat Evolution

A literature review indicated that much of the earlier work concerning the measurement of metabolic heat involved the use of crude calorimeters. Other techniques have also been used to measure the microbial heat release [42,43,44]. In most cases, the

Table 2. Values of $Y_{x/0}$ of various microorganisms growing aerobically in minimal media (presumably without producing noncellular products)

Microorganism	Substrate	$Y_{x/0}$ (g g^{-1})
Aerobacter aerogenes [33]	Maltose	1.50
Aerobacter aerogenes [33]	Mannitol	1.18
Aerobacter aerogenes [33]	Fructose	0.42
Aerobacter aerogenes [33]	Glucose	0.40
Candida utilis [34]	Glucose	1.32
Penicillium chrysogenum [35]	Glucose	1.35
Pseudomonas fluorescens [34]	Glucose	0.85
Rhodopseudomonas spheroides [34]	Glucose	1.46
Saccharomyces cerevisiae [36]	Glucose	0.97
Aerobacter aerogenes [33]	Ribose	0.98
Aerobacter aerogenes [33]	Succinate	0.62
Aerobacter aerogenes [33]	Glycerol	0.97
Aerobacter aerogenes [33]	Lactate	0.37
Aerobacter aerogenes [33]	Pyruvate	0.48
Aerobacter aerogenes [33]	Acetate	0.31
Candida utilis [34]	Acetate	0.70
Pseudomonas fluorescens [34]	Acetate	0.46
Candida utilis [34]	Ethanol	0.61
Pseudomonas fluorescens [34]	Ethanol	0.42
Klebsiella sp. [38]	Methanol	0.56
Methylomonas sp. [39]	Methanol	0.53
Pseudomonas sp. [40]	Methane	0.20
Pseudomonas sp. [41]	Methane	0.19
Pseudomonas methanica [41]	Methane	0.17

techniques require relatively complicated apparatus and/or procedures for determining the heat produced during growth and product formation.

4.1 Microcalorimetry

The literature contains a very large number of descriptions of calorimeters. General principles of design and operation of reaction calorimeters have been discussed by Skinner et al. [45]. In this review, discussion is restricted to a few representative instruments satisfactory for microbial studies. It is worthy of mention that microcalorimetry is the study of small heat changes, not necessarily with small quantities of material.

The biggest task in the investigation of microbial reaction by calorimeters is to maintain instrumental stability for long periods. In order to achieve this objective, calorimeters should operate on the twin calorimeter system introduced by Joule. Measurements are made by comparison of the temperature or some function of it between two calorimeters. As a result, long-term drifts can be neglected since both biorector vessels are equally affected.

4.1.1 Adiabatic Calorimeters

Adiabatic calorimeters consist of one reaction vessel and one balance vessel. These vessels are made as nearly identical as possible and contained within the same jacket.

The vessels are well insulated from their environment to minimize the experimental error due to heat leakage and the maximum possible temperature rise is produced in the reaction vessel. The temperature sensors are usually thermistor or resistance thermometers.

During the experiment, electrical heating is applied to the matching balance vessel so that it maintains the same temperature as the reaction vessel. Heating is also arranged to operate continuously through a feedback control circuit so that any heat produced by microbial metabolism in the reaction vessel is continuously balanced by electrical heating of the reference balance vessel. The quantity of electrical heating is continuously recorded but temperature is not measured directly.

In adiabatic calorimeters, the electrical power supplied is calculated as follows:

$$Q = \int_0^t \frac{V^2}{R} dt \tag{19}$$

where

V = voltage applied to the heater (reference balance vessel)
R = resistance of the electrical heater

The absolute sensitivity of the adiabatic calorimeter has been reported to be 2.1 $J\,h^{-1}$ and this device is usually designed to handle a small liquid volume (250 cm^3) [46]. This is a comparatively elaborate device and the calculation of the integral in Eq. (19) is a somewhat bothersome task. It is also impractical to withraw samples from the reaction vessel during the course of an experiment. As a result, parallel incubation is necessary if sampling is to be carried out. The application of adiabatic calorimeters is limited to experiments in batch culture.

4.1.2 Thermal Fluxmeter

Unlike adiabatic calorimeters where the reaction vessel is isolated, a thermal flux-meter allows heat to flow along a controlled path to a heat sink. Heat flow rate is detected by monitoring the temperature gradient along this path. The measuring element in such calorimeters is a multiple-junction thermocouple (up to 10,000 electrolytically formed junctions have been employed) and the thermopile is also the controlled path through which heat passes. It is worthy of mention that when the junctions of the thermopile are maintained at different temperatures, an e.m.f. (electromotive force) is produced. Conversely, when a current is applied to a thermo-pile one set of junctions becomes hotter and the other set cooler. The direction and the magnitude of the temperature gradient depends on the polarity and the magnitude of the impressed current (the Peltier effect).

Similar to the adiabatic calorimeter, the twin calorimeter is employed and the measuring element of the reaction vessel is compared with that of an identical vessel. The design and operation of the thermal fluxmeter has been extensively described by its developers [47,48,49,50].

The absolute sensitivity of the thermal fluxmeter, which is designed to handle a very small sample (10 cm^3), is higher than that of the adiabatic calorimeter. About 0.042 $J\,h^{-1}$ is very easily detectable [46]. However, there is some zero drift at this sensitivity level. Since the thermal fluxmeter is designed for a general application some modifications to the standard design have been found necessary for microbial studies [46]. Even though this type of instrument has been widely employed in microbial work it is a relatively difficult electrical measurement to make. Its application is also limited to experiments in batch culture.

4.1.3 Flow Calorimeter

Recently, Eriksson and Wadsö [51] and Eriksson and Holme [7] described a flow calorimeter which can be used to measure heat evolution, independent of the disturbances caused by stirring or the addition of gas, alkali or nutrients. The calorimetric cell is coupled with an external reaction vessel of unrestricted volume via a pumped flow. The heat generated in the reaction vessel passes through thermopiles surrounding the measuring cell, generating an electrical potential, which, after amplification, is recorded as a heat flow rate.

Except for the wall growth, this design appears very applicable for anaerobic continuous culture system. For aerobic batch cultures, this technique suffers from the problem of oxygen deficiency. Such a drawback may be partly overcome by introduction of oxygen to the flow. As well, flow calorimeter may also be difficult to use with filamentous organisms or non-Newtonian fluids.

In summary, calorimetry is very useful to detect a small heat change since the technique is capable of high sensitivity. In the past, conventional calorimeters are limited to experiments in batch cultures. Since the development of flow calorimeters growth studies of continuous culture system are quite possible.

4.2 Dynamic Calorimetry

A successful attempt was reported by Cooney et al. [1] based on a simple technique for measuring the rate of heat production during a microbial process by monitoring the broth temperature increase when the temperature controller was turned off. A sensitive thermistor was incorporated into one leg of a Wheatstone bridge to record the rise in temperature as function of time. The sensitivity of the circuit was such that full scale, on a 30 cm recorder, corresponded to 0.772 °C when the bioreactor was operated at 37 °C, and 0.690 °C at 30 °C.

$$Q_{acc} = \Sigma M_i C_{pi} \frac{dT}{dt} \tag{20}$$

where

$$\Sigma M_i C_{pi} = (MC_p)_{broth} + (MC_p)_{bioreactor\ jar} + (MC_p)_{stainless\ steel} \tag{21}$$

The heat accumulation measured in this manner was then corrected for heat losses and gains on the bioreactor.

$$Q_f = Q_{acc} - Q_{agi} + Q_{surr} + Q_{sens} + Q_{evp} \tag{22}$$

where

Q_f = heat evolution during the microbial process
Q_{acc} = heat accumulation in the bioreactor with no temperature control
Q_{surr} = heat loss to the surroundings
Q_{sens} = heat gained by the gas stream leaving the bioreactor with respect to the heat content of the gas stream entering
Q_{evp} = heat loss due to evaporation of water from the liquid culture
Q_{agi} = heat of agitation (mechanical mixing power input)

If the incoming gaseous stream is saturated with water at the temperature of the culture broth, Q_{sens} and Q_{evp} are negligible. Equation (22), therefore, becomes:

$$Q_f = Q_{acc} - Q_{agi} + Q_{surr} \tag{23}$$

Since the impeller rotational speed and aeration rate is kept constant during the culture growth, $(Q_{agi} - Q_{surr})$ is calibrated before inoculation at the specified agitation and aeration rates

$$Q_f = Q_{acc} - Q_{agi} + Q_{surr} = 0 \tag{24}$$

Hence

$$Q_{agi} - Q_{surr} = (Q_{acc})_{t<0} \tag{25}$$

Therefore, the metabolic heat at time t is determined by

$$(Q_f)_t = (Q_{acc})_t - (Q_{acc})_{t<0} \tag{26}$$

The overall accuracy of the dynamic calorimetric technique is reported to be —1.2%. When the quantity of heat release is around 20 MJ m^{-3} h^{-1} the accuracy of dynamic calorimetry ranges from —4.5% to 1.1%.

The use of the dynamic calorimetric technique possesses a certain advantage in its relative simplicity. The approach, however, requires continuous attention during the experiment and may not be compatible with the application of automatic process control. This represents a serious disadvantage since microbial experiments or production runs can last for several days. The use of dynamic calorimetry represents a further experimental disadvantage. The temperature fluctuations resulting from the application of dynamic calorimetric technique may disturb the microbial activities. Furthermore, it usually takes a long time to obtain the heat evolution data by this approach. An initial period of 5–9 minutes is required to let the temperature rise reach a constant rate after the control system has been turned off. Dynamic calorimetry, therefore, is not as efficient in reliably determining the heat evolution during the end of the exponential culture growth phase when the rate of heat release may drastically decline in a relatively short period of time or during other rapid metabolic changes.

4.3 Continuous Calorimetry

A special temperature control system was developed and applied to continuous measuring of the heat evolved during a microbial process [52]. In this technique, the culture broth is "overcooled" by a given constant cooling water flow rate. The excess heat removed from the bioreactor is made up by the action of an immersion electrical heater. The temperature of the culture broth is measured by a thermocouple and compared with the set-point temperature of an electrical proportional controller. The difference between these two signals is employed to control the programmable direct current power supply. The heat removed from the bioreactor can be evaluated as the difference between the heat transferred from the fermentation broth to the cooling water and the compensating heat produced by the immersion heater.

Mathematically, the rates of heat removal can be expressed as follows:

$$Q_{con} = \frac{UA \times K_\alpha F_c}{V_L(K_\alpha F_c + 1)}(T - T_c) - \frac{VI}{V_L} \tag{27}$$

where

$$K' = 2C_c\varrho_c/UA \tag{28}$$

UA = specific heat transfer coefficient for cooling surfaces
F_c = flow rate of cooling water
V = voltage of immersion heater
I = current of immersion heater
T = temperature of culture broth
T_c = inlet temperature of cooling water

Considering the exothermic growth and respiration of the microorganism, the following heat balance can be written on the bioreactor:

$$Q_f = Q_{con} + Q_{surr} - Q_{agi} - Q_{bub} \tag{29}$$

$(Q_{agi} + Q_{bub} - Q_{surr})$ is calibrated before inoculation at the specified agitation and aeration rates:

$$Q_{agi} + Q_{bub} - Q_{surr} = (Q_{con})_{t<0} \tag{30}$$

Therefore, the metabolic heat at time t is determined by

$$(Q_f)_t = (Q_{con})_t - (Q_{con})_{t<0} \tag{31}$$

The overall accuracy of the continuous calorimetric technique is reported to be 2.04 ± 1.31%. The accuracy of continuous calorimetry ranges between 4.9% to .2% when the quantity of heat release varies from 12.2 MJ m^{-3} h^{-1} to 50 MJ m^{-3} h^{-1}. The use of the continuous calorimetric technique possesses some experimental advantages when compared to the dynamic calorimetric technique. Speed of response to the culture metabolic activity and its automatic operation are among the main ones.

5 Growth Field based on Total Energy, Y_{kcal}

A growth yield based on total energy available from the medium is defined as follows:

$$Y_{kcal} = \frac{\Delta X}{(-\Delta H_c)(\Delta X) + \Delta Q} \tag{32}$$

The denominator of Eq. (32) means that total energy available consists of two parts, i.e., energy incorporated biosynthetically into cellular materials and that expended by catabolism. The concept of Y_{kcal} is of great interest in microbial cultivation since it relates to the amount of energy removed from the medium during growth.

If there is no assimilation of carbon dioxide which would be brought from the air, the following relationship can be derived:

$$(-\Delta S_c) = \frac{\sigma_b}{\sigma_s} \Delta X \tag{33}$$

where

$(-\Delta S_c)$ = amount of substrate equivalent to cellular carbon produced
σ_b = weight part of carbon in biomass
σ_s = weight part of carbon in substrate

It is obvious that the difference between $(-\Delta S)$ and $(-\Delta S_c)$ is the amount of substrate that might be dissimilated in energy yielding processes:

$$(-\Delta S) - (-\Delta S_c) = (-\Delta S)\left(1 - \frac{\sigma_b}{\sigma_s} Y_{x/s}\right) \tag{34}$$

The quantity of heat evolution (ΔQ) thus can be expressed as follows:

$$\Delta Q = (-\Delta H_s)(-\Delta S)\left(1 - \frac{\sigma_b}{\sigma_s} Y_{x/s}\right) - \sum (-\Delta H_{pi})(\Delta P_i) \tag{35}$$

From Eqs. (32) and (35), it follows that:

$$Y_{kcal} = \frac{Y_{x/s}}{(-\Delta H_c) Y_{x/s} + (-\Delta H_s)\left(1 - \frac{\sigma_b}{\sigma_s} Y_{x/s}\right) - \sum (-\Delta H_{pi})(\Delta P_i)} \tag{36}$$

If any end-products are not produced, Eq. (36) might be reduced as follows:

$$Y_{kcal} = \frac{Y_{x/s}}{(-\Delta H_c)(Y_{x/s}) + (-\Delta H_s)\left(1 - \frac{\sigma_b}{\sigma_s} Y_{x/s}\right)} \tag{37}$$

Table 3. Y_{kcal} values of heterotrophs growing aerobically in minimal media (assuming no non-cellular products are produced) [24]

Microorganism	Substrate	Y_{kcal} (g kJ^{-1})
Aerobacter aerogenes	Maltose	0.0248
Candida utilis	Glucose	0.0311
Penicillium chrysogenum	Glucose	0.0256
Pseudomonas fluorescens	Glucose	0.0229
Rhodopseudomonas spheroides	Glucose	0.0268
Saccharomyces cerevisiae	Glucose	0.0294
Aerobacter aerogenes	Glucose	0.0241
		Average = 0.0277
Aerobacter aerogenes	Ribose	0.0213
	Succinate	0.0174
	Glycerol	0.0258
	Glycerol	0.0258
	Lactate	0.0119
	Pyruvate	0.0141
		Average = 0.0203
Aerobacter aerogenes	Acetate	0.0115
Candida utilis	Acetate	0.0220
Pseudomonas fluorescens	Acetate	0.0179
		Average = 0.0184
Candida utilis		0.0268
Pseudomonas fluorescens		0.0184
		Average = 0.0215
Methylomonas methanolica	Methanol	0.0256
Klebsiella sp.	Methanol	0.0194
Pseudomonas sp.	Methanol	0.0210
Pseudomonas sp.	Methane	0.0184
Methylococcus sp.	Methane	0.0248
Pseudomonas sp.	Methane	0.0129
Pseudomonas methanica	Methane	0.0119
		Average = 0.0158

Eq. (37) is used to calculate Y_{kcal} (σ_b = 0.5 and $(-\Delta H_c)$ = 22.17 kJ g^{-1} are assumed).

The values of Y_{kcal} of heterotrophs, growing aerobically in minimal media without producing noncellular products are summarized in Table 3 [24]. In general, the average values of Y_{kcal} in the case of glucose appears to be higher than those of other substrates, particularly in the cases of C_1 and C_2 compounds. This could be due to the fact that when a low or high molecular substrate is transported into the cells, more energy (ATP) is expended for the entry of a low molecular weight substrate on the basis of the same mass of substrate consumed [24].

6 Relationship between Heat and Mass Balance

When growing on ammonia as the only source of nitrogen, microorganisms utilize different organic substrates, according to the following equation:

$$CH_mO_l + a\,NH_3 + b\,O_2 \rightarrow y\,CH_pO_nN_q + z\,CH_rO_sN_t + c\,H_2O + d\,CO_2 \quad (38)$$

where

CH_mO_l = elemental composition of the organic substrate
$CH_pO_nN_q$ = elemental composition of the biomass
$CH_rO_sN_t$ = elemental composition of the extracellular products

Respective reductance degrees for substrate, product, and cells which are the equivalents of available electron (av. e) per atom of C are defined as follows:

$$\gamma_s = 4 + m - 2l \quad (39)$$
$$\gamma_p = 4 + r - 2s - 3t \quad (40)$$
$$\gamma_b = 4 + p - 2n - 3q \quad (41)$$

Where the number of equivalents of available electrons is taken as 4 for carbon, 1 for hydrogen, —2 for oxygen and —3 for nitrogen. Therefore, there are no available electrons in CO_2, H_2O, and NH_3.

When the type of substrate or product is specified, the values of γ_s and γ_p can be directly calculated from Eqs. (39) and (40). The value of γ_b can also be calculated from Eq. (41) when the elemental composition of the biomass is identified.

The values of γ_b for various species of microorganisms grown on various substrates under various cultivation conditions are given in Table 4.

Based on Eq. (38), an electron balance can be expressed as

$$\gamma_s + (-4)\,b = y\,\gamma_b + z\,\gamma_p \quad (42)$$

Table 4. Reducing power (γ_b) of dried biomass of some microbial species [2)]

Microorganism	Carbon source	γ_b
Candida tropicalis	n-alkanes	4.385
Bacterium	n-pentane	4.607
Pseudomonas	n-alkanes	4.497
Candida sp.	n-alkanes	4.260
Saccharomyces cerevisiae	Glucose[a]	4.237
Saccharomyces cerevisiae	Glucose	4.291
Saccharomyces cerevisiae	Ethanol	4.469
Saccharomyces cerevisiae	Acetic acid	4.416
Pseudomonas aeruginosa	n-hexane	4.320
Candida sp.	n-alkanes	4.278
Candida sp.	Glucose	4.074
Candida sp.	Cellulose hydrolyzate	4.110
Candida sp.	Cellulose hydrolyzate	4.122
Candida sp.	Cellulose hydrolyzate	4.201
Candida sp.	Cellulose hydrolyzate	4.087
Candida sp.	Cellulose hydrolyzate	4.220
Candida sp.	Cellulose hydrolyzate	4.214
Candida sp.	Cellulose hydrolyzate	4.199
Candida sp.	Cellulose hydrolyzate	4.143
Candida sp.	Cellulose hydrolyzate	4.212

[a] Anaerobic growth. The rest of the data are for aerobic growth

Hence

$$\frac{4b}{\gamma_s} + \frac{\gamma_b}{\gamma_s} y + \frac{\gamma_p}{\gamma_s} z = 1 \quad (43)$$

Equation (43) can also be rewritten in the form

$$\frac{4Q_0 b}{Q_0 \gamma_s} + \frac{Q_0 \gamma_b}{Q_0 \gamma_s} y + \frac{Q_0 \gamma_p}{Q_0 \gamma_s} z = 1 \quad (44)$$

Where Q_0 is the amount of heat evolved per equivalent electrons transferred to oxygen.

According to Solomon and Erickson [53], Q_0 is related to the quantity of energy which is incorporated into biomass through biosynthetic process as follows:

$$Q_0 = \frac{12(-\Delta H_c)}{\sigma_b \gamma_b} - 13.91 \quad (45)$$

However, in Eq. (45) a correction has been made for the biomass nitrogen which is combusted to N_2 instead of NH_3.
It is obvious from Eqs. (43) and (44) that:

$$\varepsilon = \delta_h = 4 \frac{b}{\gamma_s} \quad (46)$$

$$\eta = \varepsilon_g = \frac{\gamma_b}{\gamma_s} y \quad (47)$$

$$\xi_p = \varepsilon_p = \frac{\gamma_p}{\gamma_s} z \quad (48)$$

where

η = fraction of available electrons in the substrate which is incorporated into biomass
ε = fraction of available electrons in the substrate which is transferred to oxygen
ξ_p = fraction of available electrons in the substrate utilized in product formation
γ_h = fraction of substrate energy evolved as heat
ε_g = fraction of substrate energy converted to biomass
ε_p = fraction of substrate energy converted to product

It is obvious from Eq. (44) that, under aerobic conditions, the fraction of the substrate energy evolved as heat (γ_h) is equal to the fraction of available electrons transferred to oxygen (ε).

6.1 Relationship among η, $Y_{x/s}$, $Y_{av.e}$, and Y_{kcal}

From the definition of y (Eq. 38)

$$y = \frac{\Delta X}{(-\Delta S)} \frac{\sigma_b}{\sigma_s} \tag{49}$$

$$y = Y_{x/s} \frac{\sigma_b}{\sigma_s} \tag{50}$$

From Eqs. (47) and (50), the following equation is derived:

$$\eta = Y_{x/s} \left(\frac{\sigma_b}{\sigma_s}\right) \left(\frac{\gamma_b}{\gamma_s}\right) \tag{51}$$

Growth yield $Y_{av.e}$ is defined as follows [29]:

$$Y_{av.e} = \frac{Y_{x/s}}{Y_{av.e/s}} \tag{52}$$

Where $Y_{av.e/s}$ is electron available from substrate, av.e/mole.

Based upon the number of moles of oxygen required for the perfect combustion of one mole of substrate, $Y_{av.e/s}$ can be calculated by means of multiplying the amount of oxygen required for combustion by four, i.e., the number of electrons required for the reduction of one molecule of oxygen. For instance, $Y_{av.e/s}$ of glucose is calculated to be 24 av.e since 6 moles of oxygen are required for the combustion of 1 mole of glucose, and 1 mole of oxygen corresponds to 4 equivalents of electron, i.e., 4 av.e. The average value of $Y_{av.e}$ from 79 microorganisms calculated by Payne [30] was 3.07 gcell/gav.e and carbon sources of 69 samples were C_4 to C_6 compounds. The values of $Y_{av.e}$ from C_1 to C_3 compounds are significantly lower than those of C_6 compounds. Again, this could be explained by the fact that when a low or high molecular weight substrate is transported into the cells, more energy (ATP) is expended for the entry of a low molecular weight substrate on the basis of the same mass of substrate consumed [24]. For example, one mole of ATP is required for the transport of 180 g of glucose (1 mole) while 3 moles of ATP are required for the transport of 180 g of acetate (3 moles).

From Eqs. (51) and (52), it follows that:

$$\eta = (\sigma_b \gamma_b)(Y_{av.e}) \frac{Y_{av.e/s}}{\sigma_s, \gamma_s} \tag{53}$$

From the definitions of σ_s, γ_s, and $Y_{av.e/s}$:

$$\sigma_s \gamma_s = 12 Y_{av.e/s} \tag{54}$$

It is noted that 1 g atom of carbon is equal to 12 g mass of carbon.

Therefore:

$$\eta = \frac{\sigma_b \gamma_b}{12} Y_{av.e} \tag{55}$$

From the definition of Y_{kcal}, the following equation can be derived:

$$Y_{kcal} = \frac{(\Delta X/-\Delta S)}{[(-\Delta H_c)\,\Delta X + \Delta Q]/(-\Delta S)} \tag{56}$$

Hence

$$Y_{kcal} = \frac{Y_{x/s}}{[(-\Delta H_c)\,\Delta X + \Delta Q]/(-\Delta S)} \tag{57}$$

If no end-products are produced, the denominator of Eq. (57) can be defined as the ratio between kcal substrate consumed per gram of szbstrate consumed.

$$E_c = \frac{(-\Delta H_c)\,\Delta X + \Delta Q}{(-\Delta S)} \tag{58}$$

From Eqs. (51), (57), and (58), the relationship between η and Y_{kcal} can be expressed as follows:

$$\eta = \frac{\sigma_b\,\gamma_b}{\sigma_s\,\gamma_s} E_c\,Y_{kcal} \tag{59}$$

However, if:

$$E_c^* = \frac{\text{kcal substrate consumed}}{\text{av.e substrate consumed}} \tag{60}$$

is available, then:

$$\eta = \frac{E_c^* \sigma_b \gamma_b}{12} Y_{kcal} \tag{61}$$

It is noted that:

$$E_c = \frac{\sigma_b \gamma_b}{12} E_c^* \tag{62}$$

From Eqs. (51), (55), and (59), the following general equation is established:

$$\eta = Y_{x/s} \frac{\sigma_b\,\gamma_b}{\sigma_s\,\gamma_s} = \frac{\sigma_b \gamma_b}{12} Y_{av.e} = \frac{\sigma_b \gamma_b}{\sigma_s \gamma_s} E_c Y_{kcal} \tag{63}$$

6.2 Relationships among Heat Evolution Data, Oxygen Consumption Data, Carbon Dioxide Respiration Data, and Substrate Consumption Data

It has been reported that some substrate consumption could occur in the absence of growth or product formation. This observation leads to the concept of maintenance requirement [54, 55], i.e., growing microorganisms require a certain amount of energy for maintenance purposes. This energy is required to support cell viability, cell motility, enzyme turnover, osmotic pressure, nutrient storage, and other processes referred to as cell maintenance functions.

Marr et al. [54] and Pirt [55] have identified some equations which relate growth yields and the specific growth rate (μ):

$$q_s = m_s + \frac{1}{Y_{x/s}^{max}} \mu + \frac{1}{Y_{p/s}^{max}} \frac{1}{X} \left(\frac{\Delta P}{\Delta t}\right) \tag{64}$$

$$q_{O_2} = m_O + \frac{1}{Y_{x/O}^{max}} \mu + \frac{1}{Y_{p/O}^{max}} \frac{1}{X} \left(\frac{\Delta P}{\Delta t}\right) \tag{65}$$

$$q_{CO_2} = m_D + \frac{1}{Y_{x/D}^{max}} \mu + \frac{1}{Y_{p/D}^{max}} \frac{1}{X} \left(\frac{\Delta P}{\Delta t}\right) \tag{66}$$

$$q_h = m_Q + \frac{1}{Y_{Q}^{max}} \mu + \frac{1}{Y_{p/Q}^{max}} \frac{1}{X} \left(\frac{\Delta P}{\Delta t}\right) \tag{67}$$

where

q_s = rate of substrate utilized per gram cells
q_{O_2} = rate of oxygen consumption per gram cells
q_{CO_2} = rate of carbon dioxide evolution per gram cells
q_h = rate of heat evolution per gram cells
m_s = rate of organic substrate consumption for maintenance
m_O = rate of oxygen uptake for maintenance
m_D = rate of carbon dioxide evolution for maintenance
m_Q = rate of heat evolution for maintenance
$Y_{x/s}^{max}$ = biomass "true" yield based on organic substrate
$Y_{x/O}^{max}$ = biomass "true" yield based on oxygen consumption
$Y_{x/D}^{max}$ = biomass "true" yield based on CO_2 evolution
biomass "true" yield based on heat evolution
μ = specific growth rate

Assuming that the relationship between ($\Delta P/\Delta t$) and ($\Delta X/\Delta t$) can be related as follows:

$$\frac{1}{X}\frac{\Delta P}{\Delta t} = \alpha + \beta \left[\frac{1}{X}\frac{\Delta X}{\Delta t}\right] \tag{68}$$

or

$$\frac{1}{X}\frac{\Delta P}{\Delta t} = \alpha + \beta\mu \tag{69}$$

From Eqs. (64) to (67) and (69), it follows that:

$$q_s = m_s + \frac{1}{Y_{x/s}^{max}} \mu + \frac{1}{Y_{p/s}^{max}} (\alpha + \beta\mu) \tag{70}$$

$$q_{O_2} = m_O + \frac{1}{Y_{x/O}^{max}} \mu + \frac{1}{Y_{p/O}^{max}} (\alpha + \beta\mu) \tag{71}$$

$$q_{CO_2} = m_D + \frac{1}{Y_{x/D}^{max}} \mu + \frac{1}{Y_{p/D}^{max}} (\alpha + \beta\mu) \tag{72}$$

$$q_h = m_Q + \frac{1}{Y_Q^{max}} \mu + \frac{1}{Y_{p/Q}^{max}} (\alpha + \beta\mu) \tag{73}$$

From substrate utilization data

Introducing the rate of consumption of available electrons in organic substrate per available electron in biomass, δ, gives:

$$\delta = \frac{\sigma_s \gamma_s}{\sigma_b \gamma_b} \left[\frac{-\Delta S}{X \, \Delta t} \right] \tag{74}$$

From Eqs. (70) and (74), the following relationship can be established:

$$\delta = m_s \frac{\sigma_s \gamma_s}{\sigma_b \gamma_b} + \frac{1}{Y_{x/s}^{max}} \frac{\sigma_s \gamma_s}{\sigma_b \gamma_b} \mu + \frac{1}{Y_{p/s}^{max}} \frac{\sigma_s \gamma_s}{\sigma_b \gamma_b} (\alpha + \beta\mu) \tag{75}$$

Similar to Eq. (51), η^{max} is related to $Y_{x/s}^{max}$ as below:

$$\eta^{max} = Y_{x/s}^{max} \left[\frac{\sigma_b \gamma_b}{\sigma_s \gamma_s} \right] \tag{76}$$

and

$$m_e = m_s \frac{\sigma_s \gamma_s}{\sigma_b \gamma_b} \tag{77}$$

Where m_e is defined as the energetic maintenance

Eq. (75), therefore, can be rewritten as follows:

$$\delta = m_e + \frac{1}{\eta^{max}} \mu + \frac{1}{Y_{p/s}^{max}} \frac{\sigma_s \gamma_s}{\sigma_p \gamma_p} \left[\frac{\alpha \sigma_p \gamma_p}{\sigma_b \gamma_b} + \frac{\beta \sigma_p \gamma_p}{\sigma_b \gamma_b} \mu \right] \tag{78}$$

It is noted that:

$$\xi_\varrho^{max} = Y_{p/s}^{max} \frac{\sigma_p \gamma_p}{\sigma_s \gamma_s} \tag{79}$$

and

$$\alpha_e = \frac{\sigma_p \gamma_p}{\sigma_b \gamma_b} \alpha \tag{80}$$

$$\beta_e = \frac{\sigma_p \gamma_p}{\sigma_b \gamma_b} \beta \tag{81}$$

Eq. (78), therefore, becomes:

$$\delta = m_e + \frac{1}{\eta^{max}} \mu + \frac{1}{\xi_p^{max}} (\alpha_e + \beta_e \mu) \tag{82}$$

From the definition of δ:

$$\delta = \delta_g + \delta_m + \delta_p \tag{83}$$

where

δ_g = the specific rate of consumption of organic substrate because of growth

δ_m = the specific rate of consumption of organic substrate because of cell maintenance

δ_p = the specific rate of consumption of organic substrate because of product formation

The fraction of energy in the organic substrate that is evolved as heat, δ_h or ε, is also expressed as follows:

$$\varepsilon = \varepsilon_g + \varepsilon_m + \varepsilon_p \tag{84}$$

where

ε_g = the fraction of energy in organic substrate that is evolved as heat because of growth

ε_m = the fraction of energy in organic substrate that is evolved as heat because of maintenance

ε_p = the fraction of energy in organic substrate that is evolved as heat because of product formation

Upon the above considerations, the following equations can be derived:

$$\eta^{max} = \frac{\eta}{\eta + \varepsilon_g} \tag{85}$$

and

$$\xi_p^{max} = \frac{\xi_p}{\xi_p + \varepsilon_p} \tag{86}$$

where

$$\xi_p = Y_{p/s} \frac{\sigma_p \gamma_p}{\sigma_s \gamma_s} \tag{87}$$

From Eqs. (78) and (83), it is obvious that:

$$\frac{\delta_g}{\delta} = \frac{1}{\eta^{max}} \frac{\mu}{\delta} \tag{88}$$

and

$$\frac{\delta_m}{\delta} = \frac{m_e}{\delta} = \varepsilon_m \tag{89}$$

It is further noted that:

$$\frac{\mu}{\gamma} = \frac{\sigma_b \gamma_b}{\sigma_s \gamma_s} Y_{x/s} = \eta \tag{90}$$

therefore,

$$\frac{\delta_g}{\delta} = \frac{\eta}{\eta^{max}} = \eta + \varepsilon_g \tag{91}$$

and

$$\varepsilon_m = \frac{m_e}{\delta} = \frac{\delta_m}{\delta} = m_e \frac{\eta}{\mu} \tag{92}$$

Similarly, the product-associated fraction of energy in organic substrate that is evolved as heat is expressed as follows:

$$\frac{\delta_p}{\delta} = \xi_p + \varepsilon_p \tag{93}$$

From Eqs. (86) and (93), it follows that:

$$\frac{\delta_p}{\delta} = \xi_p + \varepsilon_p = \frac{\xi_p}{\xi_p^{max}} \tag{94}$$

Eq. (83) can be rewritten as below:

$$1 = \frac{\eta}{\eta^{max}} + m_e \frac{\eta}{\mu} + \frac{\xi_p}{\xi_p^{max}} \tag{95}$$

Eq. (95), therefore, can be rearranged as follows:

$$\frac{\mu}{\eta}\left[1 - \frac{\xi_p}{\xi_p^{max}}\right] = m_e + \frac{1}{\eta^{max}}\mu \tag{96}$$

From oxygen consumption data
From the definition of δ:

$$\delta = \frac{\sigma_s\gamma_s}{\sigma_b\gamma_b}\left[\frac{1}{X}\frac{-\Delta S}{\Delta t}\right] = \frac{\sigma_s\gamma_s}{\sigma_b\gamma_b}\left[\frac{1}{X}\left(\frac{-\Delta O_2}{\Delta t}\right)\left(\frac{-\Delta S}{-\Delta O_2}\right)\right] \tag{97}$$

or

$$\delta = \frac{\sigma_s\gamma_s}{\sigma_b\gamma_b}\left[\frac{-\Delta S}{-\Delta O_2}\right] q_{O_2} \tag{98}$$

From the stoichiometric balance of Eq. (38):

$$\frac{-\Delta O_2}{-\Delta S} = \frac{\sigma_s b}{12} \tag{99}$$

Therefore,

$$\delta = \frac{\sigma_s\gamma_s}{\sigma_s\gamma_s}\left(\frac{12}{\sigma_s b}q_{O_2}\right) = \frac{48}{\varepsilon\gamma_b\sigma_b}q_{O_2} \tag{100}$$

or

$$\delta\varepsilon = \frac{48}{\sigma_b\gamma_b}q_{O_2} \tag{101}$$

$\delta\varepsilon$ is defined as the equivalents of oxygen consumed per available electron in biomass.

Eq. (83) is rewritten as follows:

$$\delta\varepsilon = \delta\varepsilon_g + \delta\varepsilon_m + \delta\varepsilon_p \tag{102}$$

It is noted that:

$$\varepsilon_g = \frac{\mu}{\delta\eta^{max}}(1 - \eta^{max}) \tag{103}$$

$$\varepsilon_p = \frac{\xi_p}{\xi_p^{max}}(1 - \varepsilon_p^{max}) \tag{104}$$

From Eqs. (82), (102), (103) and (104), the following equation is established:

$$\delta\varepsilon = \frac{\mu}{\eta^{max}}(1 - \eta^{max}) + m_e + \frac{(\alpha_e + \beta_e\mu)}{\xi_p^{max}}(1 - \xi_p^{max}) \tag{105}$$

From Eqs. (71) and (101), the following relationship can be derived:

$$\delta\varepsilon = \frac{48}{\sigma_b\gamma_b} m_0 + \frac{48}{\sigma_b\gamma_b}\left(\frac{\mu}{Y_{x/0}^{max}}\right) + \frac{48}{\sigma_b\gamma_b Y_{p/0}^{max}}(\alpha + \beta\mu) \tag{106}$$

Comparing Eqs. (105) and (106):

$$m_e = \frac{48}{\sigma_b\gamma_b} m_0 \tag{107}$$

$$\frac{1}{Y_{x/0}^{max}} = \frac{\sigma_b\gamma_b}{48}\left(\frac{1}{\eta^{max}} - 1\right) \tag{108}$$

$$\frac{1}{Y_{p/0}^{max}} = \frac{\gamma_p\sigma_p}{48}\left(\frac{1}{\xi_p^{max}} - 1\right) \tag{109}$$

From heat evolution data

In aerobic processes, heat evolutions can be directly related to oxygen utilization as discussed previously. It is further noted that 1 mole of oxygen corresponds to 4 equivalents of an electron, i.e. the number of electrons required for the reduction of one molecule of oxygen:

$$q_h = 4Q_0 q_{O_2} \tag{110}$$

From Eqs. (101) and (110):

$$\delta\varepsilon = \frac{12}{\sigma_b\gamma_b}\left[\frac{q_h}{Q_0}\right] \tag{111}$$

Therefore, the following equation can be derived from Eqs. (73) and (101):

$$\delta\varepsilon = \frac{12}{\sigma_b\gamma_b}\left(\frac{q_h}{Q_0}\right) = \frac{12}{\sigma_b\gamma_b}\left(\frac{m_Q}{Q_0}\right) + \frac{12}{\sigma_b\gamma_b Q_0}\left(\frac{\mu}{Y_Q^{max}}\right) + \frac{12(\alpha + \beta\mu)}{\sigma_b\gamma_b Q_0 Y_{p/Q}^{max}} \tag{112}$$

It is obvious from Eqs. (105) and (112) that:

$$m_e = \frac{12}{\sigma_b\gamma_b}\frac{m_Q}{Q_0} \tag{113}$$

$$\frac{1}{Y_Q^{max}} = \frac{\sigma_b\gamma_b Q_0}{12}\left[\frac{1}{\eta^{max}} - 1\right] \tag{114}$$

$$\frac{1}{Y_{P/Q}^{max}} = \frac{\sigma_b \gamma_b Q_0}{12} \left[\frac{1}{\xi_p^{max}} - 1 \right] \tag{115}$$

From carbon dioxide production data
The term δ can be rewritten as follows:

$$\delta = \frac{\sigma_s \gamma_s}{\sigma_b \gamma_b} \left[\frac{1}{X} \frac{\Delta CO_2}{\Delta t} \right] \left[\frac{-\Delta S}{\Delta CO_2} \right] \tag{116}$$

Hence

$$\delta = \frac{\sigma_s \gamma_s}{\sigma_b \gamma_b} \left[\frac{-\Delta S}{\Delta CO_2} q_{CO_2} \right] \tag{117}$$

From Eq. (38), the ratio between (—ΔS) and (ΔCO_2) can be expressed as follows:

$$\frac{-\Delta S}{\Delta CO_2} = \frac{12}{\sigma_s d} \tag{118}$$

It is noted that the CO_2 evolved per quantity of organic substrate containing 1 g atom carbon (12 g carbon), d, is divided into growth-associated, maintenance, and growth-associated CO_2, respectively:

$$d = d_g + d_m + d_p \tag{119}$$

From Eqs. (117) and (118), it follows that:

$$\frac{\delta d}{\gamma_s} = \frac{12}{\sigma_b \gamma_b} q_{CO_2} \tag{120}$$

where

$\frac{\delta d}{\gamma_s}$ is defined as mole CO_2 evolved/equiv. available electron in biomass

The carbon balance of Eq. (38) can be expressed as follows:

$$d = 1 - y - z \tag{121}$$

Therefore, the following relationship can be derived:

$$\frac{d_m \delta}{\gamma_s} = \frac{m_e}{\gamma_s} \quad \text{(for pure maintenance, } y = z = 0) \tag{122}$$

$$\frac{\delta d_g}{\gamma_s} = \frac{\mu(1 - y)}{\eta^{max} \gamma_s} = \frac{\mu}{\eta^{max} \gamma_s} (1 - (\gamma_s/\gamma_b)\, \eta^{max}) \tag{123}$$

and:

$$\frac{\delta d_p}{\gamma_s} = \frac{(\alpha_e + \beta_e\mu)(1 - z)}{\gamma_s \xi_p^{max}} = \frac{(\alpha_e + \beta_e\mu)}{\gamma_s \xi_p^{max}} [1 - (\gamma_s/\gamma_b)\, \xi_p^{max}] \tag{124}$$

Therefore:

$$\frac{\delta d}{\gamma_s} = \frac{\mu}{\eta^{max}\gamma_s} [1 - (\gamma_s/\gamma_b)\, \eta^{max}] + \frac{m_e}{\gamma_s} + \frac{\alpha_e + \beta_e\mu}{\gamma_s \xi_p^{max}} [1 - (\gamma_s/\gamma_p)\, \xi_p^{max}] \tag{125}$$

From Eqs. (72), (82) and (125), the following equations resulted:

$$m_e = \frac{12\gamma_s}{\sigma_b\gamma_b} m_D \tag{126}$$

$$\frac{1}{Y_D^{max}} = \frac{\sigma_b}{12}\left[\frac{\gamma_b}{\gamma_s\eta^{max}} - 1\right] \tag{127}$$

$$\frac{1}{Y_{p/D}^{max}} = \frac{\sigma_p}{12}\left[\frac{\gamma_p}{\gamma_s\xi_p^{max}} - 1\right] \tag{128}$$

Therefore, m_0, m_s, m_D, and m_Q are related to each other by the following relationship:

$$m_e = \frac{\sigma_s\gamma_s}{\sigma_b\gamma_b} m_s = \frac{48}{\sigma_b\gamma_b} m_0 = \frac{12\gamma_s}{\sigma_b\gamma_b} m_D = \frac{12}{\sigma_b\gamma_b Q_0} m_Q \tag{129}$$

Eq. (129), therefore, may be used to check the consistency of experimental measurements of the substrate, oxygen consumption, carbon dioxide respiration, and heat evolution during the experiment.

7 Some Applications of the Heat Evolution Parameter

7.1 Microbial Metabolic Heat Evolution as a Process Parameter

The removal of heat from the bulk liquid microbial broth is mandatory since it is crucial to maintain an optimal temperature for the desired microbial activity. The problem of heat removal is very important in the industrial process where the capital cost of the heat removal system always represents a significant proportion of the overall cost of plant items. Useful information on the microbial thermogenesis could be used for a proper design of a bioreactor cooling system. The concept of Y_{kcal} is obviously of great interest in microbial cultivation since it relates to the amount of energy removed from the medium during growth (Table 3).

When the quantity of microbial heat evolution is specified, this value then can be employed to estimate the rate of oxygen transfer to the liquid medium. In general, the heat evolution in an aerobic microbial process is directly proportional to the quantity of oxygen utilized. The more heat removed from the bioreactor simply indicates that more oxygen is transferred from the liquid medium to the microorganism.

The quantity of heat evolution could be used to estimate the raw material cost of a microbial process [56]. Assuming the quantity non-cellular products is negligible, from Eqs. (44), (47) and (63), the following equation can be derived to estimate the quantity of actually obtained biomass from 1 ton of substrate:

$$Y_{x/s} = \eta \frac{\sigma_s \gamma_s}{\sigma_b \gamma_b} = \delta_b \frac{\sigma_s \gamma_s}{\sigma_b \gamma_b} = (1 - \delta_h) \frac{\sigma_s \gamma_s}{\sigma_b \gamma_b} \tag{130}$$

According to Minkevich and Eroshin [2], biomass reductance γ_b is constant and equals 4.2, with a relative variance of only 2% and the weight part of carbon in the biomass, σ_b, is also constant and equals 0.46 (variance, 4%).
The quantity of raw material needed to obtain 1 ton of biomass is:

$$\frac{1}{Y_{x/s}} = \frac{2}{\sigma_s \gamma_s \eta} = \frac{2}{\delta_b \sigma_s \gamma_s} = \frac{2}{(1 - \delta_h)\,\sigma_s \gamma_s} \tag{131}$$

The costs of raw material needed to produce 1 ton of biomass is calculated as follows:

$$Z_1 = \frac{1}{Y_{x/s}} M = \frac{2}{\eta \sigma_s \gamma_s} M = \frac{2}{(1 - \delta_h)\,\sigma_s \gamma_s} M \tag{132}$$

where

M = costs per 1 ton of raw material.

It is obvious from Eq. (132) that a decrease of δ_h (i.e., an increase of η) decreases the expenses due to raw materials and vice versa.

The heat evolution quantity could also be used as a parameter for estimating the cost of the cultivation process [56].

In general, the microbial processing cost consists of the cost of cooling, aeration, mixing, water, utilities, and depreciation, etc. . . . The cost of aeration usually represent a significant portion of the overall processing cost.

The cost of cultivation of 1 ton of biomass (Z_2) is then calculated as follows:

$$Z_2 = \left[\varphi \frac{2\sigma_b \gamma_b}{3}\right] \left[\frac{1-\eta}{\eta}\right] = \left[\varphi \frac{2\sigma_b \gamma_b}{3}\right] \left[\frac{\delta_h}{1 - \delta_h}\right] \tag{133}$$

hence

$$Z_2 = 1.3 \frac{\varphi(1 - \eta)}{\eta} = 1.3 \frac{\delta_h}{1 - \delta_h} \varphi \tag{134}$$

So far as 1.3φ is a constant value for a definite bioreactor, the cultivation cost is determined by substrate energy efficiency utilization dependent upon $\delta_h/(1 - \delta_h)$ or $(1 - \eta)/\eta$.

It is obvious from Eq. (134) that a higher value of δ_h (i.e., lower value η) results in a higher cost of the microbial process.

The specific costs of raw materials and cultivation can obviously be the sum of the raw materials cost and the cultivation cost:

$$Z = Z_1 + Z_2 = \frac{2.M}{\eta \sigma_s \gamma_s} + 1.3 \frac{\varphi(1 - \eta)}{\eta} \tag{135}$$

$$Z = \frac{2}{(1 - \delta_h) \sigma_s \gamma_s} M + 1.3 \frac{\delta_h}{1 - \delta_h} \varphi \tag{136}$$

7.2 Microbial Metabolic Heat Evolution as an Analytical Tool for Checking the Consistency of Biomass and Substrate Data

Any error in experimental data can lead to erroneous conclusions about microbial physiology. It is, therefore, very important to verify the validity of experimental data. Equations (93), (44), and (121) can be used to verify the consistency of experimental data in either carbohydrate- and/or hydrocarbon-based microbial process:

$$\eta + \varepsilon + \xi_p \leqq 1 \tag{137}$$

$$y + z + d \leqq 1 \tag{138}$$

$$\delta_b + \varepsilon_p + \delta_h \leqq 1 \tag{139}$$

When $\eta + \varepsilon + \xi_p < 1$, or $y + z + d < 1$ and/or $\delta_b + \delta_p + \delta_h < 1$, then either an experimental error or unmeasured products resulted. This could also be due to the fact that the maintenance term is not taken into consideration. However, if $\eta + \varepsilon + \xi_p > 1$ or $y + z + d > 1$ and/or $\delta_b + \delta_p + \delta_h > 1$, then it is definite that an experimental error is involved in the data collecting processes.

According to Erickson [16], the 95% confidence intervals of Eqs. (137) and (138) are:

$$0.94 \leqq y + d + z \leqq 1.06 \tag{140}$$

$$0.93 \leqq \eta + \varepsilon + \xi_p \leqq 1.07 \tag{141}$$

When the non-cellular product is assumed to be negligible.

Eqs. (137), (138), and (139) might be reduced as follows:

$$\eta + \varepsilon \leqq 1 \tag{142}$$

$$y + d \leqq 1 \tag{143}$$

$$\delta_b + \delta_h \leqq 1 \tag{144}$$

The applicability of the heat evolution data for checking the consistency of the biomass production data and the glucose utilization data are illustrated in Table 5.

Table 5. The applicability of the heat evolution data for checking the consistency of other culture data

Regime of cultivation:	ε_{O_2}	ε_b	ε_p	$\varepsilon_{O_2} + \varepsilon_b + \varepsilon_p$
Fed-batch (Experiment 1)	0.452	0.578	0	1.03
Fed-batch (Experiment 2)	0.459	0.484	0	0.943
	δ_h	δ_b	ε_p	$\delta_h + \delta_b + \delta_p$
Fed-batch (Experiment 1)	0.46	0.60	0	1.06
Fed-batch (Experiment 2)	0.454	0.53	0	0.984

In Table 5, a fed-batch cultivation of *S. cerevisiae* was employed to prevent ethanol production while studying the heat generation and energetic metabolism of the micro-organisms under aerobic respiration conditions. The metabolic heat was measured independently throughout the experiment [57].

As theoretically expected, the results obtained for aerobic respiration indicated that ε_{O_2} was equal to δ_h. These fractions were almost 0.46 for both of the fed-batch experiments. A good agreement between δ_h and ε_{O_2}, therefore, confirmed the validity of the biomass data and the heat evolution data selected during the experiment.

It should be important to note that Eq. (129), which involved the heat evolution data, may be used to check the consistency of experimental measurements of the substrate utilization, the oxygen consumption, and the carbon dioxide respiration.

When the oxygen consumption, the heat evolution, the carbon dioxide respiration, and the organic substrate utilization are measured independently during the experiment, the value of m_e can be estimated from one of the above parameters. The values m_e obtained from the heat evolution data must agree with those obtained from the oxygen uptake data, the carbon dioxide production data, or the substrate utilization data. Otherwise, measurement errors are likely involved in the data collecting process (Table 6).

Table 6. Verification of experimental data by the method of mass balance [53]

Organism	Limiting factor	m_O	m_s	m_e from m_O	m_e from m_s
A. aerogenes	Glucose	1.4	.054	.04116	.04358
S. cerevisiae	Glucose	0.6	.018	.01453	.01453
E. coli	Glucose	3	.090	.07262	.07264

m_O: g mol O_2 per g dry biomass per h;
m_s: g substrate per gram of dry biomass per h;
m_e: g equiv. per g equiv. of available electrons in biomass per h or kJ per kJ biomass per h

According to Solomon and Erickson [53], the sources of error involved in data recording during a microbial process are numerous. In many cases, the conventional statistical approach fails to justify the validity of the experimental data obtained. There is therefore a need for checking the consistency of data collected. The method of heat and mass balance is very usefully, reliable and a straight forward way to verify the accuracy of data in the literature.

7.3. Indirect Estimation of Biomass Concentrations by Monitoring the Quantity of Heat Evolution

There has been considerable interest in the measurements of growth and biomass concentration during fermentation. This is extremely important for industrial processes, particularly those designed to produce high cell densities.

It is essential to stress that most efforts to design instruments capable of reliable on-line, instantaneous (or continuous) measuring of biomass concentration and growth rate have been generally unsuccessful. The most widely used techniques, based on measurements of optical properties of the cell suspension are not readily applicable to most practical fermentations due to difficulties encountered with growth on the optical surfaces, low measurement sensitivity, limited range, mycelial culture morphology, non-homogeneity of culture samples and the often high levels of colouration and suspended solids, characteristic of industrial media.

The parameter of heat produced can be used for monitoring cell growth and as a physiological parameter during the course of a microbial process.

It is assumed that only small amounts of organic products are produced, Eq. (5) can be rewritten as follows:

$$\Delta Q = \left[(-\Delta H_s)\frac{(-\Delta S)}{\Delta X} + (-\Delta H_n)\frac{(-\Delta N)}{\Delta X} - (-\Delta H_c)\right]\Delta X \qquad (145)$$

Hence

$$\Delta Q = [(-\Delta H_s)/Y_{x/s} + (-\Delta H_n)/Y_{x/n} - (-\Delta H_c)]\,\Delta X \qquad (146)$$

where

$$\frac{(-\Delta S)}{\Delta X} = \frac{1}{Y_{x/n}} \qquad (147)$$

$$\frac{(-\Delta N)}{\Delta X} = \frac{1}{Y_{x/s}} \qquad (148)$$

In a defined growth media, the terms $(-\Delta H_s)$ and $(-\Delta H_n)$ are usually known and constant. The term $(-\Delta H_s/Y_{x/s}) + (-\Delta H_n/Y_{x/n}) - (\Delta H_c)$ is constant only if the heat of combustion of biomass $(-\Delta H_c)$ remains unchanged during the experiment. It is, therefore, further assumed that the elemental composition of the microorganism remains unchanged or does not vary significantly throughout the experiment.

Upon this assumption, Eq. (146) can be rewritten as:

$$\Delta Q = K \times \Delta X \tag{149}$$

where

$$K = (-\Delta H_s)/Y_{x/s} + (-\Delta H_n)/Y_{x/n} - (-\Delta H_c) \tag{150}$$

From Eq. (150), it is obvious that the proportionality constant K is dependent upon the types of substrate and culture used in the culture experiment which is characterized by a growth-associated substrate utilization with insignificant (<3%) accumulation of extracellular metabolic byproducts.

Based on Eq. (149), Wang et al. [9], Volesky et al. [11], Luong and Volesky [10] have developed a method for using the parameter of heat produced to indirectly assess the concentration and rate of growth of cells during the experiment.

The metabolic heat parameter is viewed as a relatively reliable indicator of microbial activity. This is a rather novel approach to microbial thermogenesis. The metabolic parameter can be used as a kind of «heat probe», employed in process control and in on-line optimization of the microbial process. To follow the biomass concentration in the bioreactor, reliably and without time delay, has always been a rather bothersome task, often representing a major problem, particularly with culture broth containing suspended solids with microbial cells.

So far, the application of on-line process control in microbial processes has been very limited since the biomass concentration cannot be determined rapidly enough to be of value for control.

The applicability of Eq. (149) for estimating the biomass concentration is, of course, only valid when both extracellular products and maintenance energy terms are negligible.

When the maintenance term is taken into account and still no extracellular products are produced, Eq. (73) must be used to correlate the heat evolution data and the biomass concentration. When values of m_Q and η^{max} are known, measurement of heat evolution is sufficient for predicting the other microbial process parameters such as biomass concentration, oxygen consumption, carbon dioxide respiration and organic substrate consumption.

On the other hand, the oxygen consumption data in aerobic microbial processes can be treated in a similar way, to obtain the same information given by the heat evolution data. Zabriskie and Humphrey [58] have used the biomass yield and maintenance parameters, together with oxygen measurements, to estimate biomass production.

Under aerobic conditions, the fraction of the substrate energy evolved as heat is equal to the fraction of available electrons transferred to oxygen, i.e. either the heat evolution data or the oxygen consumption data is useful for monitoring a culture process, the choice is a matter of convenience and availability of the data collecting methods. Both the oxygen consumption rate and the heat evolution rate can now be monitored continuously during the experiment.

When two variables are measured (oxygen uptake rate and heat evolution rate) and extracellular products are expected to be present in very small amounts, known values of η^{max} and m_e (or m_Q) may be used with the oxygen consumption data and the heat evolution data to predict values for organic substrate consumption, biomass production, and CO_2 production. The predicted values obtained from the oxygen uptake data must agree with those obtained from the heat evolution data. Otherwise, extracellular products may be present in significant amounts or measurement errors are likely involved in the data collecting process. For carbohydrates, the respiratory quotient (the ratio between the carbon dioxide respiration rate and the oxygen consumption rate) is approximately 1 and the heat evolution and oxygen consumption ratio is 112.8 kJ g^{-1} equivalent oxygen consumed. Good agreement provides a check on the accuracy of these two measurements. The values of η^{max} for some important microorganisms are reported elsewhere [53].

When the quantity of extracellular products is no longer negligible, the calculation of the biomass concentration from the heat evolution data becomes more complicated since the energy allocation to product formation needs to be known in this case. Direct measurement of the quantity of products formed is usually the best way to estimate the energy allocated to product formation. Known values of η^{max}, ξ_p^{max}, and m_Q may then be used with the heat evolution data to predict the values for biomass concentration, organic substrate utilization, oxygen consumption and CO_2 respiration.

The measurements of the quantity of products formed is quite possible but a significant time lag may exist in the measurement process. It is, therefore, more desirable to indirectly calculate the quantity of products formed.

In a special case when the product formation is growth-associated, the oxygen consumption data can be used together with the heat evolution data for calculating the quantity of products. The quantity of heat evolution is first calculated from the oxygen consumption data. This value is then compared with the measured value. The difference between the measured value and the predicted value of heat evolution is the estimated heat evolution due to product formation. This value may then be used to predict the product formed.

7.4 Indirect Estimation of Glucose and Ethanol Concentration during an Ethanol Biosynthesis Process

In an ethanol biosynthesis process, the growth yield and the enthalpy change per mole of substrate consumed are constant throughout the culture. The quantity of heat evolved at any time, therefore, is proportional to the amount of substrate metabolized:

$$\Delta Q = K_s(-\Delta S) \tag{151}$$

and the rate at which the substrate is consumed by the culture is thus proportional to the rate of heat evolution:

$$Q_f = K_s\left[-\frac{dS}{dt}\right] \tag{152}$$

Since the biomass is directly proportional to the amount of substrate consumed (assuming initial biomass concentration is negligible)

$$X = Y_{x/s}(-\Delta S) \tag{153}$$

From Eqs. (151) and (153), it follows that:

$$X = \frac{Y_{x/s}}{K_s} \Delta Q \tag{154}$$

The cellular rate of catabolic activity (A_x), i.e. the rate of substrate metabolized per unit of cell is expected by the following relationship:

$$A_x = \frac{1}{X}\left[-\frac{dS}{dt}\right] \tag{155}$$

From Eqs. (154) and (155), the cellular rate of catabolic activity A_x is related to the heat evolution quantity by the following expression:

$$A_x = \frac{1}{Y_{x/s}} \frac{Q_f}{\Delta Q} \tag{156}$$

It is obvious from Eq. (156) that when the proportionality constant K_s is known, glucose concentration at any time during the microbial process can be calculated from the heat evolution data. Ethanol concentration can also be calculated from the heat evolution data since,

$$\frac{dP}{dt} = Y_{p/s} \frac{-dS}{dt} \tag{157}$$

It is important to note that with the methods now available, the determination of either glucose concentration or ethanol concentration directly in the bioreactor is not possible. Instead, samples must be withdrawn for laboratory essay.

During the course of the culture, if $(1/A_x)$ is plotted against $1/S$, the slope of this straight line represents the Michaelis-Menten constant. The intersection of this line to the ordinate axis represents a theoretical maximal A_x.

The heat evolution data, therefore, should be used for the routine assessment of microbial kinetics and the cellular metabolic activity.

From a broader and interesting application of the heat evolution parameter, this quantity can be used as another parameter for process control and on-line optimization of an ethanol biosynthesis process.

8 Conclusions

In a well-defined medium and when the products and the heat of combustion of microbial cells are specified the heat evolution quantity can be accurately estimated from the material balance data. In a complex medium and/or when non-cellular products are not completely specified, the quantity of heat evolution must be experimentally determined by either microcalorimetry or dynamic calorimetry and/or continuous calorimetry. Microcalorimetry usually possesses a very high sensitivity where a small heat change is easily detectable. This technique, however, requires relatively elaborate apparatus and/or procedures. On a bench scale, the dynamic calorimetric technique possesses a certain advantage in its simplicity. This approach, however, only provides intermittent data and requires continuous attention during the experiment. Continuous calorimetry appears applicable for continuous monitoring of heat evolution during a microbial process. If the continuous record of heat production can be interpreted in conjunction with thermodynamic data and chemical information, the method can provide an overall picture of the metabolic activity of a microbial process.

The heat evolution data, measured independently during the course of a microbial process, can be used as an analytical tool for checking the consistency of other culture data such as biomass concentration, substrate utilization, carbon dioxide production, and oxygen utilization.

Under aerobic conditions (extracellular products are negligible) when both heat evolution and oxygen consumption data are measured, the fraction of the substrate energy evolved as heat is equal to the fraction of available electrons transferred to oxygen. Furthermore if the measurements are accurate the heat evolution and oxygen consumption ratio is approximately constant and about 112.8 kJ per g equivalent O_2. The heat evolution and/or oxygen consumption data then can be used for indirectly estimating the microbial growth and biomass concentration. The predicted values obtained from the heat evolution data must agree with those obtained from the oxygen uptake data. Both the oxygen consumption rate and the heat evolution rate can now be easily, continuously and instantaneously processed for on-line control/process optimization purposes.

When some quantity of non-cellular products is present, the calculation of the biomass concentration from the heat evolution data becomes complicated and a direct measurement of the quantity of product formed is desirable. However, in some cases, the quantity of noncellular products can be predicted from the heat evolution and the oxygen consumption data. The quantity of heat evolution is first calculated from the oxygen consumption data (112.8 kJ per g equivalent O_2). This value is then compared with the measured value. The difference between the measured value and the predicted value of heat evolution may then be used to estimate the product formed. While this approach can not be expected to be very accurate, it can be adequately close to provide a useful estimation.

When oxygen consumption, CO_2 production, heat evolution, and substrate consumption are independently measured, these parameters can be used in a variety of ways to check the accuracy or consistency of experimental results. The values of m_s, m_o, m_D, and m_Q are first calculated from the experimental data. The value of m_e can then be predicted from m_s, m_o, m_D, and m_Q. If the measurements are accurate, the value of m_e obtained from the heat evolution data must agree with those obtained from the oxygen uptake data, the CO_2 production data, and the substrate comsumption data. Otherwise, the results are not consistent and this methodology can identify which set of experimental data is not accurate. Undoubtedly, a combination of heat and mass balance is a powerful and quick technique for verifying the data collected during a microbial process.

In anaerobic ethanol biosynthesis, the heat evolution rate can be continuously monitored by flow calorimetry and/or continuous calorimetry. The heat evolution data could also be useful for prediciting the cellular metabolic activity and for routine assessment of microbial kinetics.

It can be concluded, therefore, that heat evolution data together with other culture data can provide a reliable data base for predicting the cellular metabolic activity and microbial kinetics. The sources of error involved in data collection from microbial processes as they are usually reported in the literature are numerous. As well, this combined approach involves for less error is a very useful, quick, and a reliable way to obtain or to verify data which define a biosynthetic process.

9 Nomenclature

a	mol ammonia per quantity organic substrate containing 1 g atom carbon (g mol per g atom C)
A_c	amount of oxygen required for the combustion of dry cell (mol g^{-1})
A_n	amount of oxygen required for the combustion of ammonia (mol g^{-1})
A_s	amount of oxygen required for the combustion of organic substrate (mol g^{-1})
A_π	cellular rate of catabolic activity (h^{-1})
b	mol oxygen per quantity organic substrate contaning 1 g atom carbon (g mol per g atom carbon)
c	mol O_2 per quantity organic substrate containing 1 g atom carbon (g mol per g atom carbon)
C_c	heat capacity of cooling water (MJ kg^{-1} $°C^{-1}$)
C_{pi}	heat capacity (MJ kg^{-1} $°C^{-1}$)
d	mol carbon dioxide per quantity organic substrate containing 1 g atom carbon (g mol per g atom carbon)
d_g	mol growth-associated carbon dioxide evolved per quantity organic substrate containing 1 g atom carbon (g mol per g atom carbon)
d_m	mol carbon dioxide evolved for maintenance per quantity substrate containing 1 g atom carbon (g mol per g atom carbon)
d_p	mol product-associated carbon dioxide evolved per quantity organic substrate containing 1 g atom carbon (g mol per g atom carbon)
E_c	constant in Eq. (58) (kJ substrate consumed per g substrate consumed)
E_c^*	constant in Eq. (60) (kJ substrate consumed per av. e substrate consumed)
F_N	molal flow rate of inert (N_2) gas (m^3 h^{-1})
F_c	flow rate of cooling water (m^3 h^{-1})
I	current of immersion heater (A)
K	proportionality constant in Eq. (149) (kJ g^{-1})
K_s	proportionality constant in Eq. (151) (kJ g^{-1})
K_α	constant defined in Eq. (28) (h)

l	atomic ratio of oxygen to carbon in organic substrate (dimensionless)
m	atomic ratio of hydrogen to carbon in organic substrate (dimensionless)
m_D	rate of carbon dioxide evolution for maintenance (g mol per g biomass per h)
m_e	rate of organic substrate consumption for maintenance (g equiv. available electrons per g equi. available electrons in biomass per hour or kJ per kJ biomass per h)
m_0	rate of oxygen uptake for maintenance (g mol per g biomass per hour)
m_Q	rate of heat evolution for maintenance (kJ per g biomass per hour)
m_s	rate of organic substrate consumption for maintenance (g per g biomass per h)
M	cost per 1 ton of raw material ($ per ton)
M_i	mass quantity (kg m^{-3})
n	atomic ratio of oxygen to carbon in biomass (dimensionless)
N_c	numbers of carbon atoms in an empirical molecular formula for cells
N_H	number of hydrogen atoms in am empirical molecular formula for cells
N_0	number of oxygen atoms in an empirical molecular formula for cells
P	product concentration (kg m^{-3})
P_T	total pressure (atm)
p	atomic ratio of hydrogen to carbon in biomass (dimensionless)
$P^{in}_{CO_2}$ $P^{out}_{CO_2}$	partial pressure of carbon dioxide in inlet and outlet gas (atm)
$P^{in}_{O_2}$, $P^{out}_{O_2}$	partial pressure of oxygen in inlet and outlet gas (atm)
P^{in}_{w}, P^{out}_{w}	partial pressure of water in inlet and outlet gas (atm)
q	atomic ratio of nitrogen to carbon in biomass (dimensionless)
q_h	rate of heat evolution per gram cells (kJ g^{-1} h^{-1})
q_s	rate of substrate utilized per gram cells (g g^{-1} h^{-1})
$\dot{q}_{O_2}$	rate of consumption of oxygen (g mol m^{-3} h^{-1})
q_{CO_2}	rate of evolution of carbon dioxide per gram cells (g mol g^{-1} h^{-1})
q_{O_2}	rate of consumption of oxygen per gram cells (g mol g^{-1} h^{-1})
Q	electrical power supplied (kJ)
Q_{acc}	heat accumulation in the bioreactor with no temperature control (MJ m^{-3} h^{-1})
Q_{agi}	heat of agitation (MJ m^{-3} h^{-1})
Q_{bub}	heat dissipated by the bubbling gas (MJ m^{-3} h^{-1})
Q_{con}	heat removal by the action of controller (MJ m^{-3} h^{-1})
Q_{evp}	heat loss due to evaporation of water from the culture broth (MJ m^{-3} h^{-1})
Q_f	microbial metabolic heat (MJ m^{-3} h^{-1})
Q_{sens}	heat gained by the gas stream leaving the bioreactor with respect to the heat content of the gas stream entering (MJ m^{-3} h^{-1})
Q_{surr}	heat loss to the surroundings (MJ m^{-3} h^{-1})
Q_0	metabolic heat evolution (kJ per g equiv.)
r	atomic ratio of hydrogen to carbon in products (dimensionless)

R	resistance of the calorimeter heater (ohm)
s	atomic ratio of oxygen to carbon in products (dimensionless)
S	organic substrate concentration (kg m^{-3})
t	atomic ratio of nitrogen to carbon in products (dimensionless) or time (h)
T	temperature of culture broth (°C)
T_c	inlet temperature of cooling water (°C)
UA	specific heat transfer coefficient for cooling surfaces (MJ $m^{-3} h^{-1} °C^{-1}$)
V	voltage of immersion heater (V dc)
V_L	liquid volume (m^3)
X	biomass concentration (kg m^{-3})
y	biomass carbon yield (fraction of organic substrate carbon in biomass) (dimensionless)
$Y_{av.e}$	growth yield based on electron available (g per g av. e)
$Y_{av.e/s}$	total electron available from substrates (av. e per mol)
$Y_{p/D}^{max}$	product «true» yield based on carbon dioxide (g product per g mol CO_2 evolved)
$Y_{p/0}^{max}$	product «true» yield based on oxygen consumption (g product per g mol O_2 consumed)
$Y_{p/Q}^{max}$	product «true» yield based on heat evolved (g product per kJ heat evolved
$Y_{p/s}^{max}$	product «true» yield based on organic substrate consumption (g product per g substrate)
Y_{Q}^{max}	biomass "true" yield based on heat evolution (g biomass per kJ heat evolved)
$Y_{x/D}^{max}$	biomass "true" yield based on CO_2 evolution (g biomass per g mol CO_2 evolved)
$Y_{x/n}$	biomass yield on nitrogen substrate (g biomass per g nitrogen substrate)
$Y_{x/0}^{max}$	biomass "true" yield based on oxygen consumption (g biomass per g mol O_2 consumed)
$Y_{x/s}$	biomass yield on organic substrate (g biomass per g substrate)
$Y_{x/s}^{max}$	biomass "true" yield based on organic substrate (g biomass per g substrate)
z	mol product per quantity organic substrate contaning 1 g atom carbon (g mol per g atom carbon)
Z	costs of material and cultivation of 1 ton ob biomass ($ per ton)
Z_1	cost of raw material for producing 1 ton of biomass ($ per ton)
Z_2	cost of cultivation of 1 ton of biomass ($ per ton)

Greek Letters

α	rate coefficient for product formation (g product per g biomass per h)
α_e	rate coefficient for product formation (available electron product per available electron biomass per hour or kJ product per kg biomass per h)

β rate coefficient for product formation (g product per g biomass)

β_e rate coefficient for product formation (available electrons product per available electron biomass or kJ product per kJ biomass)

γ_b reductance degree of biomass (equiv. available electrons per g atom carbon)

γ_p reductance degree of products (equiv. available electrons per g atom carbon)

γ_s reductance degree of organic substrate (equiv. available electrons per g atom carbon)

δ specific rate of consumption of organic substrate (available electrons substrate per available electron biomass per h or kJ kJ^{-1} h^{-1})

δ_g specific rate of consumption of organic substrate due to growth (available electrons substrate per available electron biomass per h)

δ_h fraction of energy in organic substrate that is evolved as heat (dimensionless)

δ_m specific rate of consumption of organic substrate due to maintenance (available electrons per available electron per h)

δ_p specific rate of organic substrate consumption due to product formation (available electrons per available electron per hour)

$(-\Delta H_c)$ heat of combustion of biomass (kJ g^{-1})

$(-\Delta H_n)$ heat of combustion of ammonia (kJ g^{-1})

$(-\Delta H_o)$ heat generation based on oxygen consumed (kJ per g mol O_2)

$(-\Delta H_o^*)$ heat generation based on oxygen consumed (kJ per g mol O_2)

$(-\Delta H_{pi})$ heat of combustion of product (kJ g^{-1})

$(-\Delta H_s)$ heat of combustion of organic substrate (kJ g^{-1})

$(-\Delta N)$ amount of ammonia utilized (kg m^{-3})

$(-\Delta O_2)$ amount of oxygen utilized (g mol O_2 m^{-3})

(ΔP) amount of product produced (kg m^{-3})

(ΔP_i) amount of ith product produced (kg m^{-3})

ΔQ cummulative of metabolic heat (kJ m^{-3})

$(-\Delta S)$ amount of substrate utilized (kg m^{-3})

$(-\Delta S_c)$ amount of substrate equivalent to cellular carbon produced (kg m^{-3})

Δt culture time (h)

ΔX quantity of biomass produced (kg m^{-3})

ε fraction of available electrons in the substrate which is transferred to oxygen (dimensionless)

ε_b fraction of the substrate energy converted to biomass (dimensionless)

ε_g growth-associated fraction of energy in organic substrate that is evolved as heat (dimensionless)

ε_m fraction of energy in organic substrate that is evolved as heat because of cell maintenance (dimensionless)

ε_p product-associated fractions of energy in organic substrat that is evolved as heat (dimensionless)

η fraction of energy in organic substrate that is converted to biomass or biomass energetic yield (dimensionless)

η^{max} "true" biomass energetic yield (dimensionless)

μ specific growth rate (h^{-1})

ξ_p	fraction of energy in organic substrate that is converted to products (dimensionless)
ξ_p^{max}	"true" product energetic yield (dimensionless)
ϱ_c	density of cooling water (kg m^{-3})
σ_b	weight fraction carbon in biomass (dimensionless)
σ_c	weight fraction carbon in organic substrate (dimensionless)
σ_p	weight fraction carbon in products (dimensionless)
φ	expenditures of microbial process exploitation ($ per ton O_2)

10 References

1. Cooney, C. L. et al.: Biotech. Bioeng. *11*, 269 (1969)
2. Minkevich, I. G., Eroshin, V. K.: Folia Microbiol. *18*, 376 (1973)
3. Imanaka, T., Aiba, S.: J. Appl. Chem. Biotechnol. *26*, 559 (1976)
4. Luong, J. H. T., Volesky, B.: Can. J. Chem. Eng. *58*, 497 (1980)
5. Belaich, J. P. et al.: J. Bacteriology *95*, 1750 (1968)
6. Forrest, W. W.: Microcalorimetry. In: Methods in Microbiology (ed.) Norris, J. R., Ribbons, D. W., p. 285, Academic Press 1972
7. Eriksson, R., Holme, T.: Biotech. Bioeng. Symp. *4*, 581 (1973)
8. Mou, D. G., Cooney, C. L.: Biotech. Bioeng. *18*, 1371 (1976)
9. Wang, H. et al.: Europ. J. of Appl. Microbiol. and Biotechnol. *5*, 207 (1978)
10. Luong, J. H. T., Volesky, B.: Can. J. Chem. Eng., *60*, 163 (1982)
11. Volesky, B. et al.: J. Chem. Technol. Biotechnol. *32*, 650 (1982)
12. Minkevich, I. G., Eroshin, V. K.: Biotech. Bioeng. Symp. No. 4, 21 (1973)
13. Eroshin, V. K.: ibid. No. 4995 (1973)
14. Minkevich, I. G. et al.: Microbiol. Promishlement *2*, 144 (1977) (In Russian)
15. Minkevich, I. G., Eroshin, V. K.: Stud. Biophys. *59*, 67 (1976)
16. Erickson, L. E.: Annals N.Y. Academy of Sciences *326*, 73 (1979)
17. Erickson, L. E.: Biotech. Bioeng. *21*, 725 (1979)
18. Erickson, L. E., Solomon, B. O.: Joint US-USSR Conf. on Microbial Processes, M.I.T. Cambridge, Mass. 1970
19. Erickson, L. E. et al.: Biotech. Bioeng. *20*, 1623 (1978)
20. Erickson, L. E. et al.: ibid. *20*, 1595 (1978)
21. Erickson, L. E. et al.: ibid. *21*, 575 (1979)
22. Erickson, L. E.: Biotech. Bioeng. Symp. *9*, 49 (1979)
23. Okunuki, K.: Fermentation Chemistry, Kyoritsu Shuppan Co., Tokyo 1951
24. Nagai, S.: In: Advances in Biochemical Engineering, Vol. 11 (eds.) Ghose, T. K., Fiechter, A., Blakebrough, N., Berlin: Springer 1979
25. Prochazka, G. J. et al.: Biotech. Bioeng. *15*, 1007 (1973)
26. Himmelblau, D. M.: Basic Principles and Calculations in Chemical Engineering, Prentice Hall, NJ 1967[2]
27. Giese, A. C.: Cell Physiology, W. B. Saunders Co., Philadelphia (1961)
28. Mennett, R. H., Nakayama, T. O. M.: Appl. Microbiol. *22*, 772 (1971)
29. Mayberry, W. R. et al.: ibid. *15*, 1332 (1967)
30. Payne, W. J.: Ann. Rev. Microbiol. *24*, 17 (1970)
31. Cooney, C. L. et al.: Biotech. Bioeng. *19*, 55 (1977)
32. Mateles, R. I.: ibid. *13*, 581 (1971)
33. Hadjipetrou, L. P. et al.: J. Gen. Microb. *36*, 139 (1964)
34. Hernadez, E., Johnson, M. J.: J. Bacteriol. *94*, 996 (1967)
35. Pirt, S. J., Callow, D. S.: J. Appl. Bacteriol. *23*, 87 (1960)
36. Nishizawa, Y. et al.: J. Ferment. Technol. *52*, 526 (1974)
37. Von Meyenburg, H. K.: Arch. Microb. *66*, 289 (1969)
38. Nishio, N. et al.: J. Ferment. Technol. *55*, 151 (1977)

39. Dostalek, M., Molin, N.: In: Single Cell Protein II (eds.) Tannenbaum, S. R., Wang, D. I. C., The MIT Press 1975
40. Nagai, S. et al.: J. Appl. Chem. Biotech. *23*, 549 (1973)
41. Johnson, M. J.: Science *155*, 1515 (1967)
42. Battley, E. H.: Physiol. Plantarum *13*, 674 (1960)
43. Carlyle, R. E., Norman, A. G.: J. Bacteriology *41*, 699 (1941)
44. Forrest, W. W.: J. Sci. Instr. *38*, 143 (1961)
45. Skinner, H. A. et al.: In: Experimental Thermochemistry (ed.) Skinner, H. A., p. 157, New York: Wiley 1962
46. Forrest, W. W.: Methods Microbiol. *6B*, 285 (1972)
47. Calvet, E.: In: Experimental Thermochemistry (ed.) Skinner, H. A., p. 385, New York: Wiley 1962
48. Calvet, E.: In: Recent Progress in Microcalorimetry (ed.) Skinner, H. A., p. 1, London: Pergamon Press 1963
49. Benzinger, T. H., Kitzinger, C.: In: Temperature — Its Measurement and Control in Science and Industry, p. 43, New York: Reinhold 1963
50. Benzinger, T. H.: Fractions *2*, 1 (1965)
51. Eriksson, R., Wadsö, I.: Proc. 1st Eur. Biophys. Congr. *4*, 319 Wien Med. Akad. (1971)
52. Luong, J. H. T., Volesky, B.: J. Europ. Appl. Microbiol. Biotech. *16*, 28 (1982)
53. Solomon, B. O., Erickson, L. E.: Process Biochemistry *16*, 44 (1981)
54. Marr, A. G. et al.: Annal. NY Acad. Sci. *102* 3, 536 (1962)
55. Pirt, S. J.: Pro. Roy. Soc. *B163*, 244 (1965)
56. Eroshin, V. K.: Process Biochemistry *12*, 29 (1977)
57. Yerushalmi, L., Volesky, B.: Biotech. Bioeng. *23*, 2373 (1981)
58. Zabriskie, D. W., Humphrey, A. E.: AIChE J. *24*, 133 (1978)

Energy Balances for Ethanol as a Fuel

Federico Parisi
Istituto di Scienza e Tecnologia dell' Ingegneria Chimica dell' Università,
Via dell' Opera Pia 11, 16145 Genova, I

The energy balances relative to the production of ethanol from biomass and to its use as a fuel have been, and still are, subject of discussion and, often, object of confusion. It is proposed here to adopt definitely the SI system of units. The different balance items quoted by various authors for different raw materials are reported in such units and always referred to 1 kg of produced ethanol.

In order to compare correctly different raw materials, reasoned balances are prepared by rationalizing and normalizing the data concerning processes and products.

Once a production energy balance has been established in this way, comparison with both the production of gasoline and the respective behaviours in the engine are used as evaluation criteria. The aim is to evaluate the actual advantages in the use of ethanol as a fuel.

* "Fermentation" is used in this article as a synonym for ethanol formation.

1 Introduction

When ethanol was first considered as a possible fuel, it was claimed that several problems would emerge. These included the availability of raw materials, the reduction of agricultural areas which might lead to an increase of hunger in the world, the high production costs, and, last but not least, the energy balance. Certainly, a comparison between the 41.8 MJ net heat value of 1 kg of gasoline and the 26.7 MJ of 1 kg of ethanol would support these opinions.

Less biased, but not less interesting, data concerning the production of ethanol to be used as a fuel and its relevant energy balances were subsequently provided. As often happens, the balances furnished by those who were anxious to demonstrate the validity of their own theories did not reflect the real situation, or at least led to partial interpretations. The situation was further complicated by the difference of evaluation of single elements, the omission of some of them in the balances and, finally, the confusion coming from the use of different units (hectare and acre; kg and pound; bushel; joule, calorie and Btu; litre and gallon and maybe the further differences between American and British units and so on)[1].

It would be impossible and unnecessary to the purpose here to quote the complete literature on the subject. Data are now numerous and reliable enough and ideas clear enough to allow both a precise definition of the question and the suggestion of some proposals for the problem of the energy balance of ethanol as a fuel.

The same proposals could be suitably adapted for the energy balances of butanol-1 and acetone (or isopropanol) produced biologically and to be used as fuel, whenever the condition for their production will be more interesting than now.

Table 1 supplies conversion factors between SI units and those of different systems most frequently used in the literature.

Table 1. Conversion Factors between SI Units and Different Systems.

	To transform units into those below multiply by the corresponding factor				
	MJ kg^{-1}	kcal kg^{-1}	Btu gal.$^{-1}$	Btu USacre^{-1}	MJ kg^{-1}
Btu gal.$^{-1}$	0.35×10^{-3}	86.3×10^{-3}	—	—	—
kcal kg^{-1}	4.18×10^{-3}	—	1.80	—	—
MJ kg^{-1}	—	239.2	2.86×10^{3}	—	—
Btu USacre^{-1}	—	—	—	—	2.6×10^{-3}
MJ ha^{-1}	—	—	—	383.9	

MJ kg^{-1} may be eventually converted to MJ l^{-1} by multiplying by 0.8

1.1 What is the Meaning of an Energy Balance?

Surprisingly, some authors still strive to demonstrate that the energy balance of ethanol is negative or, at least, not very interesting, while methanol is always considered

[1] All the data will be expressed here in megajoule per kilogram (MJ kg^{-1}), always referring to 1 kg of anhydrous ethanol produced at the end of the process, unless differently specified.

a valid substitute for gasoline — at least at first, independently from costs or any other element. Indeed the energy balance of methanol from coal (or possibly from natural gas) is negative by assumption. Generally 1.75 MJ are needed to produce a quantity of methanol whose net heat value is equal to 1.00 MJ. Following [1)] the ratio methanol: coal for the heat values is 1:2. In order to obtain a better energy balance, vegetable sources should be used (e.g., wood or straw). Yet, the dimensions considered industrially valid for a methanol plant discourage such use at least in Europe.

However, attention must be paid to another fact: it is impossible to use a coal operated car without either replacing the Otto engine with a steam one or using an air gas generator. Besides, the use of natural gas as a fuel implies transportation of heavy gas cylinders on the car itself, difficulties may arise in filling, and its use is not free from risk. At present, gasoline can be partially or completely substituted only by a liquid fuel. Therefore, it is not proper to consider at the same level the energy produced from oil and the same amount of energy produced from coal, at least as far as road transport is concerned.

Another important consideration derives from this fact: for a realistic energy balance it is necessary to consider how much energy for fuel production is obtained from oil, how much generally from non-renewable energy sources (i.e., coal) and how much from a renewable source, such as a byproduct, either agricultural or industrial, of the main production (ethanol in this case).

Therefore, a merely thermic balance has only an academic meaning: a realistic view must take into account the quantity of non-renewable energy consumed and compare it with the energy produced (for automotive traction, not simply combustion energy, but mechanical energy produced).

Finally the energy produced will have to be compared with the non-renewable one resulting from the energy balance of the production of traditional fuels; it will be shown later how the total energy consumption for gasoline production and, of course, the thermal and volume efficiency of the engine will have to be taken into consideration.

2 Negative Factors in the Energy Balance

2.1 Raw Materials

A two-way distinction must be made for raw materials: those which are specifically grown for ethanol production (saccharines such as sugar beet, fodder beet, sugar cane, sweet sorghum and so on; starchy products such as potatoes, Jerusalem artichokes, cassava (manioc), grains; cellulosic materials such as wood), and those which do not require any energy consumption before collection (agricultural residues, straw, municipal wastes, etc). It is clear that the energy consumption is different in each case, therefore the two categories are treated separately.

2.1.1 Raw Materials Specifically Grown for Ethanol Production[2]

As the different balance items quoted in the following, whether inputs or outputs, will subsequently be used as terms in formulas, a progressive alphabet letter reported in parentheses has been attributed to each.

2.1.1.1 Operation of Agricultural Machinery (A)

The item with one of the highest energy consumptions is the operation of agricultural machinery (fuel and lubricating oil consumed in soil preparation, planting, cultivation, fertilizers and pesticides application, harvesting and, possibly, wood cutting). These values related to the above mentioned operations may be quite different from one kind of cultivation to another and often are quite different from place to place and from author to author.

For instance, for sugar cane production the Battelle Institute of Columbus (Ohio) indicates an average consumption of gasoline + oil + lubricants equal to 1.67 MJ kg^{-1} of ethanol [2]; a Brazilian report [3] gives 1.87 MJ kg^{-1}, but the Instituto de Fisica de Sao Paulo brings this value to 2.84 MJ kg^{-1} [4], while the same is reduced to 1.00 MJ kg^{-1} for South Africa [5].

For the ethanol production from sugar beet the Battelle Institute [2] gives 2.89 MJ kg^{-1}, while the result of an Italian study indicates that 2.01 MJ kg^{-1} are sufficient [6]. Brazilian data agree upon cassava: 1.38 MJ kg^{-1} [3] and 1.26 MJ kg^{-1} [4].

The evaluations on ethanol from corn are rather different in the United States of America: 5.02 MJ kg^{-1} according to Scheller [7] and 9.24 MJ kg^{-1} according to Chambers [9] (even if this probably includes the energy consumption for irrigation). Finally, energy consumption for sorghum, in Brazil [4], is 1.32 MJ kg^{-1}.

2.1.1.2 Irrigation (B)

Irrigation accounts for a high percentage of the energy costs, although it varies from case to case as a function of the kind of cultivation and of climatic conditions of the agricultural area.

According to the Battelle Institute [2] the average value relative to the United States for ethanol from sugar cane is 1.91 MJ kg^{-1}, while in South Africa [5] it falls to 1.3 MJ kg^{-1}. As far as sugar beet is concerned, while the Battelle Institute [2] gives a value of 3.60 MJ kg^{-1}, the Italian report [6] gives a value of 1.20 MJ kg^{-1}. For corn in the United States Scheller [7] gives a very modest value: 0.22 MJ kg^{-1}.

2.1.1.3 Chemical Products (C)

A third quite high consumption of energy is the indirect one connected with the production of fertilizers, herbicides and pesticides.

For sugar cane the average value in the United States is, according to the Battelle Institute [2], 2.95 MJ kg^{-1} of ethanol, while according to an official document [10] it is 2.30 MJ kg^{-1}; for Brazil 0.22 MJ kg^{-1} [3], 0.91 [61] or 1.14 MJ kg^{-1} [4], while in South Africa the value is 1.60 MJ kg^{-1} of ethanol [5].

---

[2] In order to have homogeneous data, where necessary and possible it is here assumed that 1 kg of ethanol is obtained from 21.0 kg of sweet sorghum; 17.0 kg of sugar cane; 13.9 kg of sugar beet; 7.0 kg of cassava and 3.5 kg of corn.

For sugar beet the Battelle Institute [2] — the values of which are generally quite high — indicates an average value of 4.20 MJ kg^{-1} of ethanol in the United States, versus the 3.5 MJ kg^{-1} of the quoted official document [10]. In Italy the value decreases to 1.50 MJ kg^{-1} [6].

In Brazil the average values relative to cassava are 0.14 MJ kg^{-1} [3] and 1.00 MJ kg^{-1} [4]; those relative to corn become very high, ranging from 4.6 MJ kg^{-1} of ethanol [4] to 7.08 [7] 7,23 [8] and up to 11.16 MJ kg^{-1} [9]; for sorghum according to [4] the value is 1.50 MJ kg^{-1}; for potatoes there is an energy consumption of 12.3 MJ kg^{-1} [11].

2.1.1.4 Machinery Production and other Energy Consumption (D)

To the factors strictly connected with the agricultural production we must however add some other elements, even if they are less evident. For instance, we must take into account that energy which is generically wasted is difficult to quantify. In sugar cane production, this kind of consumption could amount [2] to about 5% of the total of A + B + C while according to others [3,4] it would be quite negligible. And while the first author quotes 1% of the above mentioned sum in sugar beet production, the Italian project gives a value of about 6% [6]. This percentage would be of 3.3% for cassava and of 0.8% for sorghum [4], but it would increase up to 10% for corn [7,8].

However, what is too often disregarded by every information source, is the quantity of energy spent for the production of agricultural machinery. As a matter of fact, the values relative to this production should not be neglected at all. For instance, the energy consumption related to steel production is equal to 70,000 MJ t^{-1} of steel. The cultivation of sugar beets requires 750 kg of steel per hectare, corresponding to an energy consumption of 53,250 MJ ha^{-1}. Since agricultural machinery has an expected life of 10 years, the consumption corresponds to 5,325 MJ ha^{-1} y^{-1}, which is equal to 1.71 MJ kg^{-1} of produced ethanol. According to Austin [5] this value would be of 1.44 MJ kg^{-1} in the case of sugar cane and according to Chambers [9] of 2.34 MJ kg^{-1} for corn. Following Bunger [8] the figure for corn should be 2.81. As far as sugar beet is concerned, the above mentioned value is about 17.7% of the sum of values indicated in A + B + C; about 36.9% in the case of sugar cane cultivation in South Africa and about 11.5% or 21.7% in the case of corn. It is evident that the above factor is strongly influenced by the degree of mechanization in the cultivation, yet it seems correct for point 2.1.1.4 to add 20–22% to the sum, when no detailed information exists.

2.1.1.5 Transportation (E)

The agricultural product must be finally transported to the factory. This cost varies considerably as it is related to the average haul distance, and to the average capacity of the transportation vehicles (oil consumption per km per transported metric ton, possible use of rail transport). According to Battelle Institute [2] transportation has an incidence of 1 MJ kg^{-1} of ethanol for sugar cane and 2.4 MJ kg^{-1} for sugar beet. The quoted reference specifies that transportation is calculated on the following parameters: 20 km by road for sugar cane on the basis of 2.5 MJ $(t\ km)^{-1}$ and for sugar beet 16 km by road as above, and 240 km by rail on the basis of 0.44 MJ $(t\,km)^{-1}$. For sugar cane other authors suggest 0.12 MJ kg^{-1} [3], 1.05 [62] or 1.45 MJ kg^{-1} [5]; for sugar beet 1.25 [6], and 1.40 [61] MJ kg^{-1}; for cassava, a value is given of 0.12 MJ kg^{-1} of ethanol [3]; for potato culls 1.25 MJ kg^{-1} [13].

Data for corn are extremely scattered; according to Scheller [7] 0.45 MJ kg^{-1} of ethanol is the appropriate figure. According to Chambers [9] it is 2.93 MJ kg^{-1}.

The incidence of energy used for the manufacture of trucks or any other transportation vehicle on the energy consumption for transport may be considered modest, considering also the variability of other evaluations. Indeed, for amortization it has to be taken into consideration that transportation vehicles, once harvesting is completed, can be utilized for other products.

According to the Italian report [6], this consumption should not be over the 10% of the direct energy consumption for transportation.

2.1.2 Raw Materials not Specifically Grown for Ethanol Production

It is evident that, in this case, many items quoted above must not be taken into account, particularly those included in 2.1.1.2 and 2.1.1.3. Items gathered in 2.1.1.1 and 2.1.1.4 must be deeply modified and item 2.1.1.5 itself shows basic variations. Actually, the only items having some weight — indeed a relevant one — are collection and transportation.

However, their value is practically zero if collection and transportation of municipal wastes is considered (these operations must be carried out in any case and it would be quite wrong to charge ethanol production with them) and if the utilization plant is near the assembling point. When the plant for the utilization of cellulosic materials from municipal wastes is far from such location, production of ethanol will be charged with the transportation of the raw materials. However, as the production requires a preliminary concentration of the cellulosic materials, it would be advisable to perform it before transportation.

Thus, the items to be counted for municipal wastes concern the costs of concentration (separation, air classification, trommeling) and, possibly, additional transportation.

For the transportation of agricultural wastes the consumption per carried unit is generally higher than that of agricultural products since the ratio mass per volume is usually lower (as in the classic case of straw). The drawback is counterbalanced by the greater yield in ethanol of products with high cellulosic content and by the utilization of processes capable of exploiting also hemicellulosic materials. When straw is transported along an average distance of 30 km, the energy consumption would be equal to about 2.5 MJ kg^{-1} [14] or 1.2 MJ kg^{-1} of ethanol [15].

Some industrial byproducts, as molasses and whey, should be considered separately. For neither of them is, evidently, appropriate to quote energy costs for production, since they are produced anyway and thus their energy consumption is to be attributed to the industry from which they originate (sugar, dairy industry).

However, the availability of molasses is relatively modest and their best use could consist in raising the strength of too diluted worts in order to reduce subsequent energy costs in ethanol recovery. Whey availability is low and one should also consider the low yield in ethanol (about 2.5% by vol.).

2.2 Industrial Production

In order to have a realistic energy balance it is convenient to distinguish three phases: preparation of wort from the raw material, fermentation, and ethanol recovery and anhydrification. The energy consumption of each stage results from an inner energy balance of the stage itself. For instance, in the ethanol recovery stage, the energy consumption is given by the actual steam and electric energy consumption by considering all possible heat recovery, and by the energy spent for every possible operation at pressures other than atmospheric pressure. For this reason many authors quote such consumption in kg of steam and in kWh of electricity. However, these figures are not homogeneous, and, therefore, no direct addition to those of the agricultural phase is possible. Also in this case it seems more rational to express consumption in MJ, still referring to 1 kg of anhydrous ethanol produced.

2.2.1 Preparation of Wort (F)

This is surely the most variable phase of the industrial production, being strictly connected with the difference in the raw materials used. Saccharine raw materials (sugar cane, sugar beet, fodder beet, sweet sorghum) seem the easiest to treat (although it is quite different to process sugar cane and sorghum or beets). Yet they all require a considerable quantity of energy (grinding and extraction for sugar cane; slicing, diffusion or pressing for sugar beet). The Italian report [6, 12] suggests combining ethanol production with sugar production, in order to decrease costs and energy consumption and to allow sugar beet fermentation all through the year. Beets are processed together with those intended for the production of refined sugar. No purification of diffusion juices is performed; juice concentration and cooking of syrups to raw sugar is performed. 40 to 45% of the sucrose contained in the juice is separated as raw sugar (and subsequently refined). The remaining 55 to 60% of the total sucrose is left in a dark syrup which can be worked year-round.

Based on the original data, the plant produces 5.65 kg of refined sugar and 8.18 kg of sucrose in the form of a dark syrup per 100 kg of beets. For the proposed plant, this production requires 1.9 kg of fuel. If all the fuel were attributed to producing only refined sugar yielding 5.65% of the weight of beets, this consumption would amount to $100:5.65 \times 1.9 = 33.6$ kg per 100 kg of refined sugar.

An average plant producing refined sugar (and molasses) and no sucrose for ethanol, with juice purification, has a fuel utilization of approximately 30% (by weight) of the produced sugar. Therefore, the difference which pertains to sucrose in the dark syrup, is only 3.6%. The quantity of fuel used for each 100 kg of sucrose in the dark syrup is then limited to $3.6\,(5.65:8.18) = 2.48$ kg of fuel. Since 2.48 kg of fuel correspond to 103.7 MJ and 100 kg of sucrose correspond to 48 kg of ethanol: $103.7:48 = 2.16$ MJ kg^{-1} of ethanol are consumed in a mixed process. After Houben [61], sugar cane milling requires 4.29 MJ kg^{-1} of ethanol, sugar beet diffusion 3.16 MJ kg^{-1}.

After Misselhorn [11], the preparation of wort from starchy raw materials requires from 0.88 to 3.00 MJ kg^{-1} of ethanol if the continuous hydrolysis of starch is used; otherwise this value may rise to 10 MJ kg^{-1}. The value for corn is of 2.82 (Schopmeyer [16]) or 1.56 MJ kg^{-1} (Bunger [8]). An American Report [17] gives 3.8 MJ kg^{-1}, while according to Scheller [7] the value reaches 8.44 MJ kg^{-1}. Such high value can be explained only if a batch process, nowadays outdated, is used. Indeed, mean values around 2–3 MJ kg^{-1} seem more reasonable.

There are no commercial plants operating by enzymatic hydrolysis of the cellulosic materials, some ones operating by acid hydrolysis are working in USSR; according to Ghose [18] energy consumption by enzymatic processes should be in the order of 10 MJ kg^{-1} of ethanol. Other authors indicate 4.88 MJ kg^{-1} in the case of aspen wood chips, which however does not include the energy consumed to produce the enzyme. Acid hydrolysis of wood could require 18.57 MJ kg^{-1} [20]; yet, this value seems exceedingly high: other authors [19] indicate 5.94 MJ kg^{-1}. According to Bertusi [14] low temperature acid hydrolysis of straw requires no more than 7.6 MJ kg^{-1} of ethanol: energy for high temperature acid hydrolysis is reported to be 6.6 MJ kg^{-1} [15].

2.2.2 Fermentation (G)

In a classic fermentation with *Saccharomyces cerevisiae* the energy consumption would be in the order of 0.1 [17] or 0.2 MJ kg^{-1} [7,15]. Yet, according to the Italian study [6] a classic fermentative process with recycling of yeast of the Melle-Boinot type would imply a consumption of 1.25 MJ kg^{-1}. Although data are lacking, fermentation by *Zymomonas mobilis*, on which presently much is hoped, may be supposed to require not significantly different consumptions, considering also the modest amounts of energy involved.

Some authors [21,22] have proposed to operate a continuous fermentation under vacuum (residual pressure 6.8 kPa). Because of ethanol evaporation occurring as it is formed, this would keep the alcohol strength of the fermenting wort low, thus increasing the fermentation rate and the stillage concentration, but obliging to maintain a quite high vacuum in the presence of ethanol vapours and carbon dioxide. According to Maiorella [23], the increase in energy consumption for fermentation + distillation would be only of 6.2% (from 15.00 to 15.93 MJ kg^{-1} which is, by itself, quite high for the present days), but according to Ghose [24] the increase in consumption would be as high as 59%.

Fermentation using thermophilic microorganisms is presently under study for industrial applications: these microorganisms should allow to utilize the above concept at much lower vacuum. These processes, which have remarkable theoretical and possibly practical interest, will have, however, to be carefully examined from an energy point of view. In spite of all possible heat recoveries and plant insulation, the problem of working at relatively high temperatures is still there, even if saving on the mechanical energy for the vacuum. No energy data are yet available for balances.

2.2.3 Ethanol Recovery and Anhydrification (H)

This operation is, no doubt, the most expensive in terms of energy. Still many remarkable variations exist depending upon the system adopted.

Distillation is certainly the most ancient and widely used method. Making use of an entrainer, superazeotropic ethanol can be obtained. Presently there is a tendency to improve such plants resorting either to direct or indirect recompression of vapours or to the multiple effect principle, with the different columns working at different pressures.

The Italian proposal [6] suggests use of ethanol at 98.5% by wt., obtained with a single column under vacuum at 17.9 kPa. In this case, 6.27 MJ kg^{-1} would be consumed (but for ethanol at 98.5% and distilling a beer 7.5% by wt.). The same proposal suggests as an alternative the utilization of a plant with cyclohexane as entrainer for ethanol at 99.5% with a consumption of 6.83 MJ kg^{-1}, making use of differential pressures.

Another proposal suggests to work under vacuum, at only 12.6 kPa, and allows to obtain alcohol at 94% by wt. with an energy consumption of 3.1 MJ kg^{-1}, that is, about 6 MJ kg^{-1} for ethanol at 99.8% [25]. The use of diethyl ether as entrainer can reduce the energy consumption for a standard 3-column system to 7.00 MJ kg^{-1} [26]. For a Buckau-Wolf plant operating with differential pressures, a value of 7.20 MJ kg^{-1} was presented [27], which is very close to the one above. Other reports [8,15,17] give values of the same order (6.30; 6.62; 7.10 MJ kg^{-1} resp.). A relatively higher value is given [19] for the beer from aspen wood chips (7.70 MJ kg^{-1}).

It is evident how energy needs vary sensibly not only with the technique adopted, but also with the alcohol strength of the beer, being inversely related to this. Leppanen et al. [28] indicate in a graph consumptions versus ethanol concentration; other authors [29], instead, suggest the following empirical equation valid for concentrations between 0,5 and 10% by volume:

$$E = 5.63 \times (1 - 0.1^{c} + c^{0,3})^{-1} \quad (1)$$

where

E is the energy required, in MJ kg^{-1}
c is the beer concentration expressed by volume.

If the most recent improvements in distillation technology are taken into account, it seems more appropriate to modify the coefficient of Eq. (1) from 5.63 to 4.80.

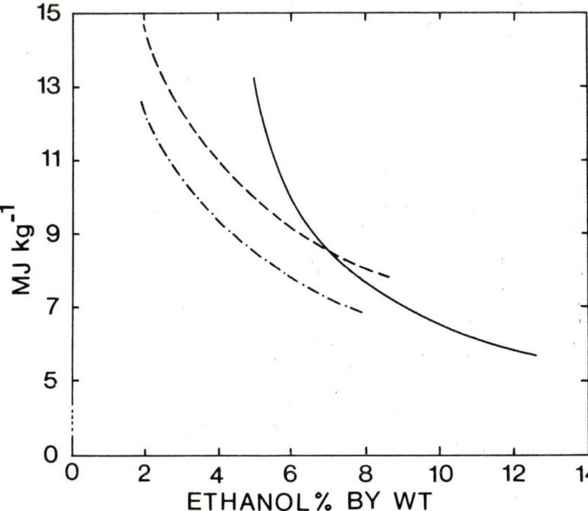

Fig. 1. Energy consumed to produce ethanol 95% by wt, vs. beer strength. ——— computed after [28]; — — — computed after [29]; —·—·— computed after expression (1) modified as in the text

Fig. 1 shows the graph following Leppanen, recalculated in MJ kg^{-1}, and the curves obtained through the equation above.

A conventional plant [11] should consume 12.5–16.0 MJ kg^{-1} but Scheller [7] quotes a consumption equal to 29.02 MJ kg^{-1}.

A process employing a third substance (entrainer) for superazeotropic ethanol will also take into account for energy consumption the normal losses of that substance (computed on the basis of its net heat value).

It is evident that, for highly diluted mashes, it may be more convenient to concentrate the mash to a suitable strength by multiple effect evaporation before fermenting it, rather than fermenting and distilling it as it is. The convenience of this operation is a matter of energetic and economic considerations and the concentration before fermentation should be adopted when stillage has to be subsequently concentrated, rather than destined to biogas production [57].

A different method may consist in the extraction of ethanol by means of solvents. Its usefulness is particularly evident in the continuous extraction during fermentation to keep the alcohol percentage low in the fermenting wort.

No energy consumption data are available for this process which is applicable only if using as solvent alcohols from C_{12} upwards, thus avoiding any toxicity for the yeast. Data are available for the extraction by means of gasoline of an alcohol brought to 90% by distillation: in this way, a 10% ethanol fuel is directly obtained, with energy savings of about 30% [30].

Other non conventional methods may consist also in producing ethanol at 85% by wt. by distillation (consumption of 2.16 MJ kg^{-1}) and subsequently dehydrating it on adsorbing materials [31], or by using molecular sieves [32] or, lastly, increasing the strength of a fermented wort by the use of membranes [33]. However, data on the energy consumption are lacking for these processes which in any case are not yet commercially operating. Moreover, membranes (which are, though, interesting also to keep the alcohol level low during the fermentation) would be easily soiled by the macromolecules in the wort. In any case their energy savings are expected to be of about 50%.

The preevaporation process [34] seems to be extremely promising, by utilizing membranes in the vapour phase, bringing ethanol from 80 to 99.8%. Starting (by distillation) from a beer at 7.5% by wt. the consumption would be of only 4.5 MJ kg^{-1}.

2.2.4 Global Data (I)

The literature on the topic often quotes final data, without details concerning adopted technologies. This procedure should be avoided, since no comparison by sectors is possible and, furthermore, it does not permit explanation of the differences between cases. Sometimes data include the combustion of byproducts, which makes the global operation even less comprehensible. For instance, for corn fermentation and distillation the U.S. Gasohol Study Group [1] quotes a consumption of 24.07 MJ kg^{-1} of ethanol. This is substantially in agreement with Scheller [7]. Yet, again for corn, Chambers [9] gives 15.0 MJ kg^{-1}, while Schopmeyer [16] indicates 6.37 MJ kg^{-1}, which agrees better with the above quotes. The information from the Gasohol Group [1] is questioned by the Comptroller General of the U.S. [35].

Concerning ethanol production from sugar cane, values range from 5.2 MJ kg^{-1} [36]

(very low indeed) to 14.17 MJ kg^{-1} of ethanol [4], but the latter is likely to include wort preparation. Austin [5] gives for sugar cane in South Africa values ranging from 41.5 to 23.36 MJ kg^{-1}. Serra [4] provides 23.32 MJ kg^{-1} for sweet sorghum and 31.10 MJ kg^{-1} for cassava, which very likely include wort preparation too.

Bertusi [14] gives for fermentation and distillation of worts from cellulosic raw materials a quite low energy consumption: 6.93 MJ kg^{-1}.

Industries producing distillation plants give generically for worts from starchy materials values from 8 to 12 MJ kg^{-1} of ethanol without further specifications. For potato culls quite different values are given from 6.25 (Böhler-Vogelbush and Vulcan Co. Inc.) to 10.00 MJ kg^{-1} for the ACR process.

Fermentation of whey and distillation require 15.57 MJ kg^{-1} [37]. To this figure 1.56 MJ kg^{-1} should be added, corresponding to the energy spent in plant construction.

Final data, as expressed above, are not very significant, because (it is worth repeating it) they do not allow in-depth distinctions of the advantages and disadvantages of the single phases as they are carried out. A high value of the sum of the energy balances (from wort preparation up to distillation) may either mean that the raw material is inadequate, or that all single operations are badly managed or badly evaluated, or that a single one has a particular deficiency, which could be improved through the introduction of a different technology. Some high energy consumptions are, in fact, peculiar features of a particular kind of process. In the direct process of sugar cane (and not of molasses!), rather diluted juices, containing 12–13% of sugar, are obtained and processed by fermentation. The distillation of quite diluted fermented worts, in spite of every possible recovery of heat, surely requires a higher energy consumption than one on worts at high concentrations. See for instance some examples [6,15,17] for which energy consumption may be evaluated at around 7 MJ kg^{-1} with beers at 9–9.4 G.L. (while worts from sugar cane should be around 7–7.5 G.L.), and Fig. 1.

For this reason it is not surprising that, according to Serra [4] ethanol from sweet sorghum requires more energy than the one from sugar cane, since the sugar juices are diluted 20% more (even if recent reports on sweet sorghum hybrids seem to deny this greater dilution). The high figure for the production from whey is surely due to the high dilution of the beer (2.5% by vol.).

The high cost of ethanol from cassava must be attributed [4] to the higher energy cost of wort preparation. The higher energy consumption in the distillation of ethanol from cellulosic materials is to be attributed to the high dilution of worts as well. Apparently, the technologies adopted in Brazil and South Africa which give the values quoted by Serra [4] and Austin [5] are not of the most advanced kind, especially as far as wort preparation and distillation are concerned. Perhaps the local availability of the fuel necessary for the process (in the case of sugar cane) blunts the sharp problem of the search for a process with low energy consumption. The opposite happens when cereals, sugar beet or potatoes are involved.

However, it is suggested that data be expressed in the form pointed out in 2.2.1; 2.2.2; 2.2.3, quoting where possible the technology adopted and avoiding any global final form.

On the mixed process "fermentation under vacuum — distillation" it has been referred already in 2.2.2.

2.2.5 Other Elements of Energy Cost (J)

In the industrial phase these elements are mainly the energy necessary to the build-up of the plant (generically, with reference to the weight of steel) and to its maintenance. The amortization period, which must be always indicated, has to be taken into account. Generally, a twenty-year amortization may be considered reasonable. According to Serra [4] this energy consumption is in the order of 2.32 MJ kg^{-1} of ethanol from sugar cane; of 2.53 MJ kg^{-1} if from sweet sorghum, and 3.3 MJ kg^{-1} if from cassava. The last value coincides with the one provided by Lindeman and Rocchiccioli [38], which assumes a higher cost in capital of the plants for the processing of cassava in comparison with those from sugar cane. These authors base their study on a twenty-years amortization and calculate that the maintenance has an incidence on the three above values of 36%, 29% and 14% respectively. According to Chambers [9], for a corn processing plant the energy amortization alone (based on the cost of the plant and a coefficient of 10.55 MJ $\$^{-1}$) would be 0.98 MJ kg^{-1} of ethanol. For a plant performing the high temperature acid hydrolysis of corn stalks the value 2.00 MJ kg^{-1} has been computed [15], although it seems slightly optimistic, considering the high corrosion phenomena. Finally, small additional consumptions have to be considered: lighting, offices, etc., which are to be charged to the industrial management and which may be approximately evaluated around 5% of the total factory needs. No cost for transportation of the final product is to be considered (except for particular cases), as it is comparable to that of gasoline.

Additional Note — Among the negative terms, the energy cost of human labour in agriculture and in the factory should also be considered. This may result in different values, according to the interpretation of socio-economic factors. An Italian evaluation [6] suggests that globally 25 full-time workers (8 hours daily for 300 days a year) are necessary to produce 1000 t of ethanol per year. If an energy demand of 12.50 MJ d^{-1} is attributed to each worker, the corresponding value is 0.11 MJ kg^{-1} of ethanol. Such value, in the author's opinion, may be disregarded in the evaluation of a balance of such a kind of accuracy.

2.2.6 Electric Power (K)

As said above, some energy balances indicate the electric power consumption as a separate item, especially when this last is expressed in kWh and thermal energy is given as quantity of steam used. The use of megajoule should allow to avoid this practice, which is to be particularly discouraged since for the major industrial plants the electric power is self-generated. However, values are not negligible. For the agricultural production of corn, for instance, the datum of 2.07 MJ kg^{-1} is supplied [8], which seems, frankly, quite high and not applicable to European production. For the industrial treatment of corn the supplied value is 4.30 MJ kg^{-1} of ethanol [17], which seems quite high too, but in any case explains the low energy consumption that the same authors indicate, for instance, for fermentation: 0,1 MJ kg^{-1}.

3 Positive Factors in the Energy Balance

3.1 Ethanol (L)

A net heat value of 26.75 MJ kg^{-1} can be assigned to ethanol. It will be shown later how this value should be interpreted for comparison with gasoline. However, it is still valid as starting point for any further consideration of an energy balance.

3.2 Agricultural Refuse and other Vegetable Products (M)

Different kinds of agricultural refuse or vegetable products may be obtained as a function of the raw materials and of the kind of process used. Their further utilization must be included in the energy balance. This item, though, does not include by-products of industrial process (like bagasse), which belong to another category. For instance Scheller [7] suggests that in a plant for ethanol from corn, 75% of the stalks, cobs and husks, be utilized. Their weight is equal to 2.58 kg kg^{-1} of ethanol. If the net heat values of these residues is 13.9 MJ kg^{-1} of residues [39], they should yield 35.9 MJ kg^{-1} of ethanol or, according to Scheller [7], up to 43.4 MJ kg^{-1}. Actually, considering their performance in the furnace, they would not produce more than 22.5 MJ kg^{-1} of ethanol [2].

From the above mentioned figure, Scheller [7] properly subtracts the energy consumption relative to the transportation of this special kind of fuel, which is probably somewhat optimistically evaluated at 0.42 MJ kg^{-1} of ethanol. It is still necessary to take into account the transportation of the possible vegetable fuel, as is done by de Carvalho [3] which charges a cassava plant with 0.2 MJ kg^{-1} of ethanol for transportation of the wood necessary to the power plant.

3.3 Utilization of By-Products (N)

The positive elements arising from the utilization of by-products may be quite varied. Above all they often are the result of a partial energy balance and are chosen according to considerations in addition to energy. For instance, is it more convenient to use beet pulps as a fuel or as a feed? From an economic viewpoint and taken case by case, is it more convenient to produce biogas from distillery bottom residues or, when soluble, to concentrate them and use them as fertilizers? Or is it preferable to separate, dry and use as feed those bottom residues which are not totally soluble? Decisions about the utilization of by-products must be made according to circumstances and economies and, of course, they may modify the energy balance. When an energy balance not referred to a specific case must be outlined, it seems more proper to always consider the best energy utilization of by-products. Indeed, the program for the production of ethanol as a fuel is energy based. Its economic factors are continuously changing to the advantage of ethanol but they cannot be precisely foreseen for the future. By-products utilizations different from the most favorable from the energy viewpoint are justified only when some specific factors become more important than the energy factors.

However, utilization with different purposes should be always specified and justified case by case. When, for instance, by-products are utilized as feed, their energy value as such should be quoted in the balance. In other cases, the energy supply could also be disregarded as for 25% of stalks, cobs and husks of corn, of which Scheller [7] rightly suggests using only 75%, leaving the remaining 25% on the ground for conditioning and trace element replacement.

Due to the large variety of cases, only the most important and significant by-products are quoted; others may be inferred by analogy.

3.3.1 Bagasse

Bagasse as obtained contains about 50% humidity. When dry, it amounts to 12% of the weight of sugar cane. One kg of dry bagasse has a net heat value of 16–17 MJ; therefore, it should yield 30–33 MJ kg^{-1} of produced ethanol, if used as an energy source. Taking into account water to be evaporated and the performance in the furnace, which decreases with lower heat values of the fuel, no more than 16–18 MJ kg^{-1} can be considered available. According to Yates [40] this value is too high and it should be brought down to 10 MJ kg^{-1}. It is interesting to note that, according to the same source, the energy consumption in the industrial phase of the production of ethanol from sugar cane should not exceed 66% of this value. Therefore either a surplus of energy should be available or the cellulose obtained from the surplus of sugar cane can be hydrolyzed.

To this purpose it is interesting to make a comparison with the ethanol which could be produced if all the available bagasse would be completely used as cellulosic raw material for production of glucose (through hydrolysis) and, subsequently, of ethanol. One ton of damp bagasse should theoretically produce 0.24 t of cellulose and 0.27 t of glucose, but practically, the glucose would not exceed 0.10–0.15 t. Consequently, the production of ethanol would be of 0.050–0.075 t. However, the ratio between damp bagasse and ethanol from sugar cane is 4 to 1 and the above mentioned data (18 and 10 MJ from the combustion of bagasse per kg of produced ethanol) reveal that, when burnt in a boiler, 1 kg of damp bagasse actually yields 4.25–2.50 MJ of usable energy. By hydrolysis, 1 kg of damp bagasse produces only 0.050–0.075 kg of ethanol, i.e., in energy terms, 1.8–2.0 MJ. Therefore, it is surely more economical to use bagasse in the production of steam and electricity. Its surplus may be devoted either to the production of the energy necessary to obtain ethanol from other sources, which are not energy self-sufficient, as cassava [40]. Similar calculations can be made for the case of sweet sorghum, for which the average dry matter of bagasse is 2.7 kg per kg of ethanol [41].

3.3.2 Beet Pulp

According to the Italian papers [6] the dry matter of pulp is 4.6% by weight of fresh sugar beets. Pulp drying can be performed in a fluidized bed multiple effect evaporator (carrier fluid: fuel oil), and 4.6 MJ kg^{-1} of dry matter are used in this operation. The heat of combustion of pulp is evaluated as 12.5 MJ kg^{-1} of pulp, 92% dry matter, that is equal to 13.6 MJ kg^{-1} of dry matter. Substracting from this value the energy for drying, the heat of combustion that can be theoretically recovered is 13.6–4.6 = 9 MJ kg^{-1} of dry matter. Since the amount of dry matter from beets corresponding to 1 kg of ethanol is 0.64 kg, 5.8 MJ kg^{-1} of ethanol is the amount of energy that can be theoretically recovered from dry pulp combustion. Since the furnace performance is lower when the fuel has a low heat combustion the available energy is estimated to be 67% of the theoretical amount. So 3.9 MJ kg^{-1} of ethanol is considered to be the amount of energy that can be effectively recovered from pulp. If the beet pulp were to be used as feed, its value could be conservatively estimated at around 13.6 MJ kg^{-1} of dry matter, i.e., 9 MJ kg^{-1} of pulp at 92% after drying (transport fluid: soy oil) with an energy value or 5.7 MJ kg^{-1} of ethanol.

3.3.3 Fermentation and Distillation Residues

The fermentation residues of cereals, of cassava and of potatoes may be dried and can constitute, together with the yeast produced, a feed of remarkable value. The value of the above residues is strongly underestimated if it is evaluated as 13 MJ kg^{-1} of residue (because of their high proteic content). The energy value of such residues must be properly recognized in any case, whether they are burnt or used as protein and energy contribution to the cattle. According to Misselhorn [11], drying of distillation residues from cereals and potatoes may be carried out with a consumption of 10 MJ kg^{-1}. Scheller [7] indicates 6.00 and Houben [61] 9.19 MJ kg^{-1} of ethanol. Yet, the energy performance of the above residues as digestible energy is at least equal to 14.9 MJ kg^{-1} of ethanol, from which the energy for the drying process must be substracted, giving roughly 5–9 MJ kg^{-1}. The figure of 6 MJ kg^{-1} is given for residues from potato culls [13].

The value 14.9 MJ kg^{-1} is in agreement with the previous one (13 MJ kg^{-1} of residues), considering that in the present case, the average dry residues yield is slightly lower than the ethanol yield. The Italian report [6] estimates especially the production of yeast: 3% in weight of yeast, 92% d.m., in comparison with ethanol. The thermal energy for drying must be entered in the balance. A consumption of 17 MJ kg^{-1} of yeast has been calculated. This amounts to 0.51 MJ kg^{-1} of ethanol. The dry yeast value in Scandinavian units is 0.98 with respect to corn (for which the energy value is 31.9 MJ kg^{-1}). Thus, the energy value of dry yeast is 0.94 MJ kg^{-1} of ethanol and the net gain is 0.43 MJ kg^{-1}.

Completely soluble bottom residues may be treated by anaerobic digestion for the production of biogas. According to the same report [6] a BOD_5 of 640,000 mg of O_2 per kg of ethanol produced (12.8 l of stillage) is obtained. Finally, 0.26 Nm^3 of biogas, having a net heat value of 25 MJ Nm^{-3}, are produced from the above. The biogas energy production is given as $25 \times 0.26 = 6.50$ MJ kg^{-1} of ethanol produced. If the waste water enters the tank at 45 C (output temperature at the distillation plant) and even if the tank is thermally insulated, some additional heat has to be supplied to keep the temperature of the fermenting liquid at 35 C. The relative energy consumption amounts to roughly 10% of the biogas produced, on a year round average. Under the above circumstances, the energy gain is about 4.70 MJ kg^{-1} of ethanol, if the energy consumed for the construction of the plant, together with other small amounts for the maintenance of the factory, is taken into account.

Following Gerletti [15], from the stillage from corn stalks it is possible to obtain 6.9 MJ kg^{-1}, 13% of which are consumed for heating the tanks.

Data provided by Misselhorn [11] give 0.250 Nm^3 of biogas with a net heat value of 24 MJ Nm^{-3}; the global energy contribution is evaluated as 6–10 MJ kg^{-1} of ethanol produced.

Whether it is convenient to produce biogas is a question requiring economic considerations, according to which other methods of utilization of the bottom residues may prevail. Such problems of choice often occur in ethanol production and in the utilization of the corresponding by-products. The reasons for any choice not aimed at the most favorable energy balance should be clearly explained. Some authors [6] believe, for instance, that in the case of ethanol produced from beet it is worth concentrating the bottom residues and using them as a trace element and potassium replacement for the soil or as additive for feed and in this case the heat consumption would be of 6.25 MJ kg^{-1} after Houben [61]; others suggest burning the solid residues obtained after evaporation and drying with a multiple effect fluid bed process. In this case, taking into account the energy wasted for evaporation and drying of residues, the energy gain is quite modest (about 1.25 MJ kg^{-1} of ethanol produced). Yet, it is possible to recover the potassium salts, which can be traded and returned to the soil as fertilizers. SCP production from bottom residues may be quite interesting and can be evaluated, according to the Italian report [6], in the order of magnitude of 0.1 kg per kg of ethanol, corresponding approximately to 1.3–1.5 MJ kg^{-1} of ethanol. However, the energy consumed during the production and drying must be taken into account. (The cell yield per m^3 of wort is modest and the energy consumption must be referred to the volume of the wort). These consumptions, varying with the process or with the particular microorganism utilized, may be evaluated in the order of 0.3 to 0.5 MJ kg^{-1} of ethanol. The actual yield in energy terms is therefore no more than 1.0 MJ kg^{-1} of ethanol.

It seems that it is possible to produce SCP acceptable as feed also in the biogas production; following Gerletti [15] this source of energy could be evaluated as 2.33 MJ kg^{-1} (net value after deduction of energy for drying) in the case of corn stalks hydrolysis.

3.3.4 Lignin

According to Bertusi [14], the low temperature acid hydrolysis of straw (and of other cellulosic materials) gives in a pre-treatment a 50% lignin solution which can be burnt with a total energy recovery of 8.70 MJ kg^{-1} of ethanol. Following other sources, the recovery of energy from lignin could be 22.83 MJ kg^{-1} if from corn stalks and 14.64–17.25 MJ kg^{-1} if from aspen wood chips (assuming enzymatic or acid hydrolysis resp. [19]).

3.3.5 Carbon Dioxide

It is clearly impossible to assign a heat value to carbon dioxide. Indeed, it corresponds to the third which is lost of the solar energy fixed as carbon in the raw material. However, it is possible to use such carbon dioxide for the production of algae, exploiting additional solar energy.

The mass quantity of carbon dioxide is almost equal to that of ethanol. For instance, it would be possible [6] to grow with it 0.88 kg per kg of ethanol of a blue alga, the *Spirulina maxima*.

The energy consumption (besides the solar energy, of course) for algae production would be generally of 9–10 MJ kg^{-1} of ethanol [6,43] and the energy gain of $0.88 \times 13 = 11.45$ MJ kg^{-1}, with a net profit of 1.45–2.45 MJ kg^{-1}. This would be proportionally lower if only a part of the carbon dioxide were devoted to this purpose.

At last, 40 m^2 per kg of ethanol all year round is needed for algae production.

4 A Pure Energy Balance of Fermentation Ethanol

Clearly, the above data would be sufficient for the energy balance of the ethanol not used as a fuel.

Indeed, following the indication of each of the preceding paragraphs, the result is substantially:

$$-\Delta H = (L + M + N) - (A + B + C + D + E + F + G + H + I + J + K) \quad (2)$$

A second method of evaluation may consist in referring the total energy produced to the non renewable one consumed

$$\mu = \frac{L + M + N}{A + B + C + D + E + F + G + H + I + J + K} \times 100 \quad (3)$$

Other proposals for compiling energy balances may seem more sophisticated, yet they are less interesting from a practical point of view.

When the terms in the former equations are assigned the values indicated by the different authors, as quoted at the corresponding items, a considerable disagreement in the results may be observed for the same raw material. This is due not only to the different estimates the authors give for some highly varying factors (i.e., for an agricultural product, the different cost for irrigation, the different cost of transportation to the factory, etc.) but also to the different technologies employed. For the distillation of corn mashes Scheller [7], for instance, suggests 29.02 MJ kg^{-1}, while Bunger [8] and Katzen [17] give, still for corn, values of less than a quarter of that amount, that is, about 7 MJ kg^{-1}.

In these conditions, any comparison among different raw materials becomes practically impossible, unless uniform criteria are adopted, with the only condition imposed by their respective processing.

A criterion of this kind has been adopted in Tables 2, 3 and 4, which compare saccharine, starchy raw materials and lignocellulosics respectively. In these Tables, figures in italics are either the average of the corresponding values given in the reported literature, or values chosen on the basis of a uniform criterion.

In the case of distillation, for instance, the value of 7 MJ kg^{-1} has been chosen for a beer if 7.2–7.5% by wt. (9.0–9.4% by vol.). This value has been suitably varied in the case of mashes which are more diluted because of the characteristics of the originating raw material (as, for instance, for sorghum and for the enzymatic hydrolysis of cellulose).

Furthermore, uniform criteria have been chosen for the evaluation of byproducts (13.0 MJ kg^{-1} of feed, whatever the feed may be; 0.4 MJ kg^{-1} of ethanol for fermentation yeasts; biogas in proportional quantity for all). In this way, values are really comparable, even if, obviously, different values may be introduced in single computations according to the particular local conditions (i.e. transportation for greater distances and so on).

From a comparative examination of Tables 2 and 3 it may be observed that the best energy yield is given by sugar cane, sorghum and cassava, practically at the same level. Corn gives brilliant results if 75% of the agricultural wastes are burnt to produce energy. Yet, this assumption turns out not so interesting in practice, nor has ever been adopted, due to the drawbacks it presents. Without this practice, the energy yield decreases to very low values, so much that it is more convenient to hydrolyze the

Table 2. Energy balances for saccharine raw materials (MJ kg^{-1})

	Sugar cane	Sweet sorghum	Sugar beet	Sugar beet
Inputs				
A) Agricultural machinery	*1.77* [2,3]	1.32 [4]	2.01 [6]	2.01 [6]
B) Irrigation	1.91 [2]	*1.91*	1.20 [6]	1.20 [6]
C) Chemical products	*1.04* [4,61]	1.50 [4]	1.50 [6]	1.50 [6]
D) Agr. machinery amort.	1.44 [5]	*1.34*	1.71 [6]	1.71 [6]
Others	0.27 [2]	*0.23*	0.23 [6]	0.23 [6]
E) Transportation	*0.75* [2,12]	*0.90*	1.25 [6]	1.25 [6]
Trucks amortization	*0.10*	*0.10*	0.13 [6]	*0.13*
F) Wort preparation	4.29 [61]	4.70 [61]	3.16 [61]	2.16 [6]
G) Fermentation	*0.60*	*0.65*	*0.60*[a]	0.50 [6]
H) Alcohol recovery	*7.00*	*8.00*	*9.00*[a]	*7.00*
I) Global data (F+G+H)	(14.17) [4]			
J) Plant amortization	2.32 [4]	2.53 [4]	1.80	2.30 [6]
Others	*0.70*	*0.77*	*0.55*	*0.48*
	22.19	23.95	23.12	20.47
Outputs				
L) Ethanol	26.75	26.75	26.75	26.75
M) Agricultural byproducts	—	*6.00*	—	—
N) Bagasse	*18.00*	*15.60*	—	—
Beet pulp	—	—	5.30 [6]	5.30 [6]
Yeast	*0.40*	*0.40*	*0.40*	*0.40*
Biogas	*3.70*	*3.00*	*3.70*[a]	4.70 [6]
	48.85	51.75	36.15	37.15
$-\Delta H$ (MJ kg^{-1}) =	26.66	27.80	13.03	16.68
μ (%) =	220	216	156	181

[a] Mashes from sugar beet diffusion are more diluted than those obtainable after [6]

agricultural wastes and ferment them to ethanol. However, as shown in Table 5 as well, also the yield per hectare is very low in the case of corn.

Given the values of $-\Delta H$ in Table 2, Table 5 shows the actual value of energy yield per cultivated hectare for each different raw material. For this table, also the yields per hectare are average values. Obviously, by modifying them the energy yield per hectare varies.

Table 5 shows clearly how erroneous it is to compute the energy yield on the basis of the produced ethanol and how much more rational it is to refer to the $-\Delta H$ per kg of the produced ethanol. Anyway, the very low convenience in the production from corn stands out clearly, as well as from cassava unless very large arable lands are available.

The hydrolysis of cellulose yields better energy results by acid hydrolysis rather than by the enzymatic one, because of the high dilution of the worts, as well as better results with wood than with agricultural wastes. The low temperature acid process seems to perform better than the high temperature one, though it requires a large amount of chemical reagents not totally recoverable. Their production energy should be taken into account in order to compile a more precise balance.

Table 3. Energy balances for starchy raw materials (MJ kg^{-1})

	Corn	Corn + hydrolysis of 75% of stalks etc.	Cassava
Inputs			
A) Agricultural machinery	*5.18* [7,8]	*5.18* [7,8]	*1.32* [3,4]
B) Irrigation	*0.22* [7,8]	*0.22* [7,8]	*0.50*
C) Chemical products	*7.15* [7,8]	*7.15* [7,8]	*0.57* [3,4]
D) Agr. machinery amort.	*2.57* [8,9]	*2.57* [8,9]	*0.48*
Others	0.63 [2]	0.63 [2]	0.12 [2]
E) Transportation	1.32 [7]	*2.10*	*1.00*
Trucks amortization	*0.13*	*0.21*	*0.10*
F) Wort preparation	3.00 [11]	*7.10*	3.00 [11]
G) Fermentation	*0.50*	*0.75*	*0.50*
H) Alcohol recovery	*7.00*	*10.50*	*7.00*
J) Plant amortization	*2.50*	*3.50*	3.30 [3]
Others	*0.57*	*1.05*	*0.57*
	30.77	40.96	18.46
Outputs			
L) Ethanol	26.75	40.12	26.75
M) Agricultural byproducts	22.50 [9]	—	—
N) Yeast	*0.40*	*0.60*	*0.40*
Biogas	*3.00*	6.00 [15]	*3.00*
Others	*7.00*	*8.45*	*6.00*
	59.65 (37.15)[a]	55.17	36.15
$-\Delta H$ (MJ kg^{-1}) =	28.88 (6.38)[a]	9.47	17.69
μ (%) =	194% (121%)[a]	135%	196%

[a] Data if stalks, husks etc. are left in the field. The combustion in a steam generator of those residues is not easy and has never been realized in practice

According to data provided by Allen [44], who gives information about utilization of newsprint, the energy cost of enzymatic hydrolysis should be over 1.8–2.0 MJ kg^{-1}. With the further addition of distillation of the fermented worts the value would be about 15 MJ kg^{-1} (fermentation included) and therefore quite interesting, even if no agricultural residues are used as energy source. Indeed, newsprint has in itself an energy value, which must enter the global energy balance, since it can be recycled, allowing energy savings in the paper industry. This fact must be taken into account in a general energy balance.

5 The Global Energy Balance for Ethanol Used as a Fuel

More complicated and sophisticated considerations are necessary for this topic. It is necessary to avoid interested pessimism and easy enthusiasm in discussing aspects of this subject.

Table 4. Energy balances for lignocellulosic raw materials (MJ kg^{-1})

	Wood chips		Straw, cornstalks etc.	
	acid hydrol.	enzym. hydrol.	low temp. acid hydrol.	high temp. acid hydrol.
Inputs				
A–E) Collection, transport.	*1.25*	*1.00*	*1.80* [14,15]	*1.80* [14,15]
F) Wort preparation	5.94 [19]	3.49 [19]	7.35 [14]	6.60 [15]
G) Fermentation	*0.50*	*0.70*	*0.50*	*0.50*
H) Alcohol recovery	*7.00*	*10.00*	*7.00*	*7.00*
J) Plant amortization	*3.00*	*4.20*	*3.50*	*3.00*
Others	*0.68*	*0.71*	*0.75*	*0.70*
	18.37	20.10	20.90	19.70
Outputs				
L) Ethanol	26.75	26.75	26.75	26.75
N) Lignin	17.25 [19]	14.64 [19]	8.70 [14]	*8.70*
Yeast	*0.40*	*0.40*	*0.40*	*0.40*
Biogas	*6.00*	*6.00*	*6.00*	6.00 [15]
	50.40	47.79	41.85	41.85
$-\Delta H$ (MJ kg^{-1}) =	32.03	27.69	20.95	22.15
μ (%) =	274%	238%[a]	200%	212%

[a] More energy is consumed in enzymatic hydrolysis because of the high dilution of fermented worts. Concentration may be performed before fermentation in a multiple effect evaporator, still the situations does not improve sensibly from the energy viewpoint

Table 5. Ethanol and energy yields per cultivated area

Raw material	Crop, t ha^{-1}	Ethanol, t ha^{-1}	Ethanol, GJ ha^{-1}	$-\Delta H$, GJ ha^{-1}
Sugar cane	55 –70	3.1–4.1	82.9–109.7	82.6–109.3
Sweet sorghum	55 –70	2.6–3.3	69.6– 88.3	72.2– 91.7
Sugar beet [6]	45 –55	3.2–3.9	85.6–104.3	53.4– 65.1
Corn	3.5– 4.0	1.0–1.1	26.7– 29.4	5.5– 6.0
Corn + (75% straw)	3.5– 4.0	1.5–1.6	40.1– 42.8	14.3– 15.7
Cassava	10 –15	1.4–2.1	37.4– 55.6	25.9– 38.9

5.1 The Energy Cost of Production of Gasoline, Ethanol and their Blends

The net heat value of 1 kg of gasoline may be conventionally fixed at 41.00 MJ kg^{-1}, but the consumption of a non-renewable energy source such as crude-oil in order to produce that kg of gasoline [9] is equal to 50.5 MJ kg^{-1} of gasoline, with an energy loss of 20.8%. In addition to this reference from the USA, a European study [45] suggests that added consumption ranging from 17 to 25% of non-renewable energy is needed to produce premium gasoline (Research Octane Number, R.O.N.: 96–98), depending on the amount of tetraethyl lead to be added. Finally, it is possible to produce unleaded gasoline with consumptions of non-renewable energy of 30% and more if R.O.N.

is maintained at the 96–98 level. Furthermore, in order to be fair, we must note that the average energy cost of transportation and the energy cost of plant construction has to be computed for crude oil as it is in the case of ethanol. These values may be extremely scattered due to the variability in both the distances and means of transportation and in the refinery operations. Taken as a whole they have been evaluated at around 10% of the energy costs quoted above. Yet it is very important to point out that ethanol blended gasoline raises remarkably its octane rating. Fig. 2 shows the increase of the R.O.N. and the increase in the average O.N. (Research and Motor) according to the Nebraska Test [46] and to the American Petroleum Institute [47].

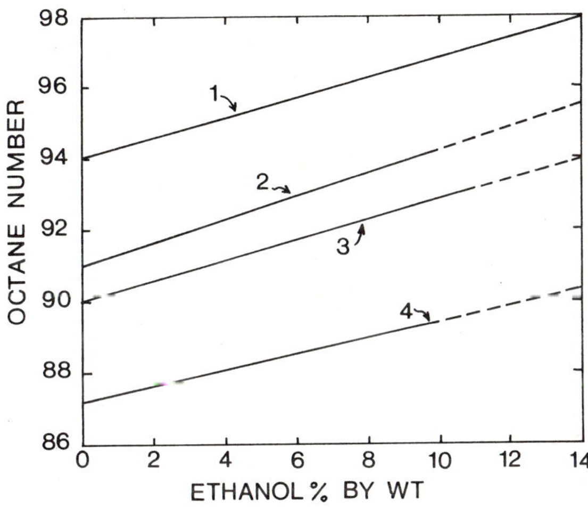

Fig. 2. Increase in the Octane Number of a gasoline due to the addition of ethanol. 1) R.O.N. after the author; 2) R.O.N. after [46]; 3) average of R.O.N. + M.O.N. after [46]; 4) average of R.O.N. + M.O.N. after [47]

The advantage given by ethanol in the octane rating increase can be quantified either in terms of savings in tetraethyl lead using the gasoline presently used (i.e., in ecological terms) or (with equal use of tetraethyl lead) starting with gasolines at lower octane ratings, which, therefore, require less energy input in their production. It is, obviously, not reasonable to expect both benefits. Still, the second method of evaluation offers the advantage of quite precise numerical data (even if social factors may suggest the first solution). After Sezzi [45], the energy savings in the production of gasoline at lower R.O.N. would generally be 2.5% of the net heat value of the fuel for gasolines blended with 10% of ethanol and 5% for gasolines blended with 20% of ethanol. After Spencer and Brandberg [48], this advantage is different depending on the different cases: however, it is 30 to 50% lower than what suggested by Sezzi.

At this point a pure energy comparison between a premium gasoline with the 10% and 20% ethanol addition respectively (ethanol having a net heat value of 21.40 MJ kg^{-1}, that is, 26.75 minus the energy expense for its production on the basis of $\mu = 180\%$), could be, after the data from Sezzi and accepting a 25% consumption in the production of premium gasoline:

$$\text{gasoline:} \quad 41.8 - (0.25 + 0.025) \times 41.8 = 30.3 \text{ MJ kg}^{-1}$$
$$\text{gasoline 9:ethanol 1:} \quad 0.9 \times [41.8 - (0.25 + 0.025 - 0.025) \times 41.8] + 0.1 \times 21.4 = 30.3 \text{ MJ kg}^{-1}$$
$$\text{gasoline 8:ethanol 2:} \quad 0.8 \times [41.8 - (0.25 + 0.025 - 0.05) \times 41.8] + 0.2 \times 21.4 = 29.5 \text{ MJ kg}^{-1}$$

According to Jawetz [49] the energy savings in the production of low quality gasoline would be 6–10%. Table 6 reports the values obtained after the different methods.

Table 6. Comparison of the energy available in premium gasoline and in lower grade gasoline-ethanol mixtures at 10% and 20% of ethanol by vol., having the same R.O.N.

Energy spent for gasoline production %	25 [45]			20.8 [9]		
Gasoline, available energy $MJ\ kg^{-1}$	30.2			32.2		
	According to			According to		
	45)	48)	49)	45)	48)	49)
Gasoline 90% + 10% ethanol	30.3	30.0–30.1	31.7–33.1	32.1	31.6–31.8	33.4–34.9
Gasoline 80% + 20% ethanol	30.2	29.4–29.7	32.5–35.2	31.7	30.8–31.3	34.1–36.6

If the different data given by a particular author [49] are disregarded, as apparently quite optimistic, the mean and the standard deviation relative to the energy of 1 kg of gasoline are 31.3 ± 0.9 MJ, having deducted all energy costs for transportation and production from crude oil. In the case of lower grade gasoline with the addition of 10% ethanol these values become 31.0 ± 0.9 MJ kg^{-1}. For mixture with 20% of ethanol, the value decreases to 30.5 ± 0.8. It can be concluded that the energy balance of gasoline production and the production of one of its blends of equal octane rating are practically equivalent, apart from small variations or those due to different evaluations. On the other hand, it is clear that if pure ethanol were used, this advantage would vanish. Ethanol may be credited at the most with the 3 MJ kg^{-1} saved in the anhydrifiçation, since, in this case, the use of hydrated ethanol under the azeotropic point is possible. The same for ethanol used for fuel cells.

Ethanol has not been given much credit as a diesel fuel because of its low Cetane Number, low density and immiscibility with diesel fuel. In any case, from an energy standpoint, its contribution is limited to the results of its energy balance [50].

5.2 Ethanol in the Engine

5.2.1 Ethanol — Gasoline Blends

When an alcohol blend is used in an unmodified engine, the most pronounced effect is a leaner fuel-air mixture [50]. An engine operating on an alcohol blend behaves as if the carburetor were adjusted to release less fuel. So, since a smaller amount of fuel is burnt, the output of the engine for any given throttle setting is correspondingly less. Peak efficiency occurs at any throttle opening when the fuel-air ratio is slightly lean; the throttle must be opened more for a lean engine to get the same power.

Since, according to the same reference [50)], the efficiency of a fuel is proportional to its net heat value, the number of km covered with 1 kg of gasoline is approximately 4% larger than the number of km covered with a blend 9:1 of gasoline + ethanol. This percentage becomes 3.6% when referred to the unit volume, because of the different densities of the two liquids.

According to Ecklund [51)] a 10% blend has a little effect upon consumption. However, Scheller [52)] reports that even a maximum advantage of 5.3% on the volume efficiency, and the advantage of about 9% on the energy would result at 7 C with a 10% blend. The advantage on the volume efficiency would decrease to zero (still according to Scheller) for temperatures around 20 C. The considerable advantage in the thermal efficiency in comparison with the MJ of the alcoholic and non-alcoholic fuel is also acknowledged by Brinkmann [53)] and was recognized by Doldi [63)] still in 1933.

After Myburgh et al. [54)] a fuel with 10% of ethanol results in a 1% higher power with respect to simple gasoline in an engine at 3,000 revolutions. Such advantage decreases to zero at 5,000 revolutions. The specific volume consumption, again at 3,000 revolutions, is 0.5% greater for the 10% mixture.

Of course, such remarks are applicable to engines with carburetor and not to the fuel injection engines.

Faced with the contrasting opinions on the relative volume efficiency (approximately expressed for blend 9:1 by values ranging from 0.96 to 1.05) as quoted above, the ACR Process Corp. [55)] suggests as a reasonable value a Salomonic 1.00, evidently considering ±5% an admissible deviation for this kind of inquiry.

5.2.2 Non-anhydrous Ethanol

Problems are only partially simplified by using non-anhydrous ethanol (95–96% by vol.), without blending it with gasoline, as it is presently practiced in Brazil. The utilization of ethanol in an Otto engine, without modifying the compression ratio, implies renouncing to an advantage most peculiar to ethanol itself, i.e., the high octane rating. Indeed the thermal efficiency increases 3% for ethanol (on a hydrous basis) in comparison with gasoline when compression ratio (C.R.) is 7.5. However, according to Brinkmann [53)] the volume efficiency is 36% lower, as suggested also by the difference between net heat values. According to the same source, using a C.R. 12 with ethanol and C.R. 7.5 with gasoline, the thermal efficiency improves up to 10%, whereas volume efficiency is reduced by only 31%. The values of C.R. greater than 12 do not appear interesting from a practical viewpoint.

Data provided by Brinkmann are comparable in automotive industry practice, but they are not homogeneous: what would the efficiency gap be between a gasoline at C.R. 7.5 or 9 (still possible in practice) or 12 (a merely hypothetical ratio) and an alcohol engine with the same compression ratios? According to the Tyzard and Pye formula, the efficiency of a gasoline engine for a C.R. 7.5; 9 and 12 for a stoichiometric ratio would be of 0.921; 0.937 and 0.956 respectively, with an improvement of 3.8%

between the two extremes. Since the experimental thermal efficiency improves by 10% with ethanol and the theoretical one with gasoline improves by only 3.8%, the advantage given by ethanol at the compression ratio 12 is clear. The value of C.R. 12 is close to the one adopted in Brazil for vehicles running on ethanol only, although it is known that such values can be somewhat exceeded.

According to Pimentel [56], the theoretical ratio (ethanol consumption/gasoline consumption) would be 1.09 in the case of ethanol at 96% (net heat value 25.1 MJ kg^{-1}), taking into account both the different efficiency and the different density. The higher actual consumption would range from 10 to 20%.

As mentioned above, the non-anhydrous ethanol also gains from the energy savings in the distillery, since the dehydration phase (for a value of 2–3 MJ kg^{-1}) is not carried out.

5.3 An Evaluation of Non-renewable Energy Savings

A paper by Ghose [18] and one by Jawetz [49] contain a calculation which can easily be put into practice with the data quoted above as follows. The first assumption is that in order to produce 1 kg of gasoline, 19% more hydrocarbons or, generally, crude oil will be consumed. It is also assumed possible to save 6% of crude oil, producing a gasoline at lower O.N., suitable for blending with 10% of ethanol. Furthermore, it is assumed that the densities of ethanol and gasoline are equal as a first approximation, that the energy yield in the ethanol production is $\mu = 200\%$ and finally, that the efficiency of ethanol blended with gasoline is equal to 1.05 (assuming the accuracy of Scheller [52]). Accepting such assumptions corresponds to admitting that 1.05 kg of gasoline is equivalent to 1 kg of ethanol in the engine and, moreover, that 2.00 kg of ethanol can be obtained from 1 kg of hydrocarbons, which means that

$$2.00 \times 1.05 = 2.10 \text{ kg of ethanol},$$

with an efficiency equivalent to 2.10 kg of gasoline, correspond to 1 kg of hydrocarbons.

It is, however, appropriate to assign to ethanol the energy savings, since it allows a saving of 6% of the hydrocarbons in the energy consumption for gasoline production. Therefore, the value 2.10 must be raised to 2.23 kg of gasoline equivalent (such as ethanol) per kg of hydrocarbons. Yet, from 1 kg of hydrocarbons, the 100% (in energy terms) of gasoline is never recovered. Even by producing low grade gasoline, 19% of the gasoline produced (expressed as hydrocarbons) will have to be consumed. Then, from 1 kg of hydrocarbons only 0.840 kg of gasoline is obtainable and, therefore, 1 kg of ethanol in fact replaces 2.65 kg of hydrocarbons. The remarkable saving of non-renewable energy oil is evident.

However, the above evaluation can be subject to some criticism. First of all, it is optimistic, since it takes into account all the highest values quoted in the literature. As already mentioned, while the energy yield in ethanol production may be prudentially evaluated (especially for future planning) on the average at around 180% instead of 200%, efficiency is a matter of debate. Similarly, the 6% energy savings in the production of poor quality gasoline may be optimistic and it should be reduced to

3%. Even under this less favorable condition and considering an efficiency of 0.97 and an energy yield for the production of ethanol of 180%, the value of 2.12 obtained is extremely positive. The real value is probably located between the latter and the former one (2.37 ± 0.15). A 20% substitution, assuming an energy yield for ethanol production of 180–200%, an efficiency of 0.92–0.95 and savings in gasoline production of only 3–5% would give a ratio 2.19 ± 0.38, which is still very interesting.

The same evaluation is considerably less favorable in the case of ethanol with no gasoline. Here the net heat value (26.75 MJ kg^{-1}) and the better efficiency have to be considered. After Brinkmann [53], 1 kg of gasoline corresponds to 1.35 kg of ethanol, which requires 0.68–0.75 kg of hydrocarbons for its production using the reference value of 200–180%. Since 1.19–1.25 kg of hydrocarbons are needed to produce 1 kg of gasoline, the ratio becomes then 1.71 ± 0.09, lower than the preceding ones, but still appealing.

5.4 Gasoline from Ethanol

Premium gasoline can be produced from ethanol using a zeolite as catalyst, by a process similar to the one devised by the Mobil Oil Corp. to transform methanol into gasoline [58]. This possibility has been confirmed by Derouane [59] in 1978 and, recently, by Tsao [60]. According to Maiorella [64], the passage from ethanol to gasoline would imply an energy loss of only 3%. It would create further energy savings of 14.3 MJ kg^{-1} considering savings in distillation and anhydrification of ethanol, as well as energy savings in not producing gasoline from crude oil.

These data seem, indeed, quite optimistic: based on the lowest consumption in distillery and considering the different densities of ethanol and gasoline, energy losses should be in the order of 7.5%. The energy savings should be of 12.4 MJ kg^{-1} from which, though, the 7.5% of 41.8 MJ kg^{-1} should be deducted, to yield a final value of 9.34 MJ kg^{-1}.

These considerations seem quite interesting, but what said in the preceding paragraphs on the real savings in energy should be kept in mind. The balance proposed by Maiorella accounts 26.75 MJ kg^{-1} for the ethanol and uses 1.74 kg of it per kg of gasoline. But, considering all the data, those of Table 6 included, 1 kg of ethanol may be considered to give in the mixture with gasoline a global yield practically equivalent to that of 1 kg of gasoline. Thus, the transformation does not seem really interesting. The Mobil Oil process maintains, on the contrary, its interest for the methanol conversion.

6 Ethanol for Chemical Uses

Ethanol may be easily converted to acetaldehyde, from which it is possible to obtain a wide range of products (e.g., acetic acid, n-butanol, etc.). It may also be converted to ethylene, which allows the production of several additional products as well. Within these processes, ethanol is also quite interesting from an energy standpoint, taking into account both the direct comparison of global energy balances and the possibility of saving crude oil to be utilized for fuel production. Since the processes may vary considerably, the energy comparisons must be made case by case and according to the various processes used. In some cases ethanol for chemical uses offered economic advantages even in 1982, but this is not the proper place to discuss this particular matter.

7 Conclusions

To summarize what has been treated thus far, it is possible to say the following:

- an energy balance has meaning only when it clearly shows the quantity of available energy in relation to the non-renewable energy consumed;
- when ethanol is used as a fuel, it is necessary to point out how much energy as liquid fuel may be gained from non-liquid fuels;
- when ethanol is used as a fuel, the energy balance must be global; as a matter of fact, it must not only take into account the production process but also the performance of the engine and the possible energy savings within the production of low grade gasoline, to be blended with ethanol;
- in any case, the energy comparison must be carried out by charging gasoline with all energy consumptions relative to its own production.

Some procedures are suggested in order to evaluate energy consumption relative to ethanol in an accurate way. Furthermore, the data from literature on this matter and the examples quoted show the clear energy advantage deriving from a partial or total substitution of ethanol for gasoline.

8 Acknowledgements

The Author wishes to thank his friend Robert F. Light, M. Sc., Prof. Richard Light, and his daughter Maria Adelaide Parisi, M. Sc., for their help.

9 References

1. U.S. Dept. of Energy: Gasohol Report of the Energy Res. Advisory Board on Gasohol, Washington, D.C. 1980
2. Battelle Inst., Columbus Labs.: System Study on Fuels from Sugarcane, Sweet Sorghum and Sugar Beets, vol. 2, Task 77, p. 166, Columbus, Ohio 1976
3. de Carvalho, A. V. Jr. et al.: Proc. Internat. Symp. on Alcohol Fuel Technology, Methanol and Ethanol, Wolfsburg, FGR 1977
4. Serra, G. E. et al.: Prepr. IIIrd Alcohol Fuels Technology Internat. Symp., vol. 2, Asilomar, Cal. 1979
5. Austin, R. B. et al.: J. Agr. Sc. *91*, 667 (1978)
6. Consiglio Nazionale delle Ricerche: Etanolo per via fermentativa, Milano 1979; Parisi, F.: Chim. Ind. (Milan) *61*, 574 (1979)
7. Scheller, Wm. A., Mohr, B. J.: Proc. of the 171st Nat. Meet. of the American Chemical Soc., vol. 2, No. 2, p. 29, New York 1976
8. Bunger, R. E.: in: Gasohol, a step to Energy Independence (Lyons, T. P., ed.), Alltech Technical Publications, Lexington 1981
9. Chambers, R. S. et al.: Science *206*, 789 (1979)

10. U.S. Federal Energy Administration and U.S. Dept. of Agriculture: Energy and U.S. Agriculture. 1974 Data Base, vol. 2, Committee Series of Energy Tables, U.S. Govn't Printing Office, Washington, D.C. 1977
11. Misselhorn, K.: Proc. of the DECHEMA Meeting, Frankfurt a/M, June 1980
12. Parisi, F.: Biological Conversion: Ethanol and Alcohol Routes. In Energy from Biomass, IInd E. C. Conf. Berlin, Sept. 20–23, 1982 (Strub, A., Chartier, P., Schleser, G., eds.), p. 803, Applied Science Publ., London, 1983; Parisi, F.: Chim. Ind. (Milan) *65*, 23 (1983)
13. Taweel, A. M. et al.: in: Advances in Biotechnology (Moo-Young, M., Robinson, C. W., eds.), vol. 2, p. 153, Oxford: Pergamon Press 1981
14. Bertusi, M., Distillerie San Marco, Ferrara, Italy: priv. comm.
15. Gerletti, M.: E. Bi. A., Broni, Italy: priv. comm.
16. Schopmeyer, D.: Alcohol for motor fuel, Archer Daniels Midland Decatur, Ill. 1972
17. Katzen, R., Associates: Grain Motor Fuel Alcohol Technical and Economic Assessment Study for the U.S. Dept. of Energy, Cincinnati, Ohio 1978
18. Ghose, T. K.: Biological production of ethanol from cane molasses and lignocellulosics in India, India Inst. of Technology Delhi, New Delhi 1980
19. Wayman, M., Lora, J. H., Gulbinas, E.: J. Am. Chem. Soc. *90*, 183 (1979)
20. Wettstein, P., De Vos, J.: Preprints of the IVth Alcohol Fuels Technology Internat. Symp., vol. 1, p. 37, Guarujà, Brazil 1980
21. Cysewski, G. R., Wilke, C. R.: Biotech. Bioeng. *19*, 1125 (1977)
22. Ramlingham, A., Fink, R. K.: ibid. *19*, 583 (1977)
23. Maiorella, B., Wilke, C. R.: ibid. *22*, 1749 (1980)
24. Ghose, T. K., Tyagy, R. D.: ibid. *21*, 1387 (1979)
25. Maiorella, B., Blanch, H., Wilke, C. R.: Lawrence Berkeley Lab. Rept. 10219, November 1979; A.I.Ch.E. 72nd Nat. Meet., San Francisco, Cal., Nov. 29, 1979
26. Wentworth, T., Othmer, D.: Trans. Am. Inst. Chem. Engnrs *36*, 785 (1980)
27. Stegemann, J.: Buckau-Wolf Informationstag, Agrotechnik, Hannover Messe 1980
28. Leppanen, O., Denslow, J., Ronkainen, P.: Preprints of the IVth Alcohol Fuels Technology Internat. Symp., vol. 1, p. 123, Guarujà, Brazil 1980
29. Wettstein, P., De Vos, J.: Preprints of the IVth Alcohol Fuels Technology Internat. Symp., vol. 1, p. 37, Guarujà, Brazil, 1980
30. Leeper, S. A., Wankat, P. C.: Ind. Eng. Chem. Proc. Dev. *21*, 331 (1982)
31. Ladisch, M., Dick, K.: Science *205*, 898 (1979)
32. Fanta, G. F. et al.: ibid. *210*, 646 (1980)
33. Pye, E. K., Humphrey, A. E.: Proc. of the IIIrd Ann. Biomass Energy System Congress, p. 69, U.S. D.O.E. Solar Energy Research Inst., Golden, Colo. 1980
34. Ballweg, A. H. et al.: Proc. of the Vth Alcohol Fuels Technology Internat. Symp., vol. 1, p. 97, Auckland, New Zealand 1982
35. Comptroller General of the U.S.: Conduct of DOE's Gasohol Study Group, Issues and Observations, EMD 80–128, Washington 1980
36. Lipinski, E. S., et al.: Fuels from Sugar Crops, Battelle Columbus Laboratories, Columbus, Ohio 1978
37. Moulin, G., Galzy, P.: in Advances in Biotechnology (Moo-Young, M., Robinson, C. W., eds.), vol. 2, p. 181, Oxford: Pergamon Press 1981
38. Lindeman, L. R., Rocchiccioli, C.: Biotech. Bioeng. *21*, 1107 (1979)
39. Miller, D. L.: Proc. of the 9th Nat. Conf. on Wheat Utilization Research, Albany, Cal. 1974
40. Yates, R. A.: Preprints of the Internat. Sweetener and Alcohol Conf., London 1980
41. McClure, T. A. et al.: Preprints of the IVth Alcohol Fuels Technology Internat. Symp., vol. 1, p. 123, Guarujà, Brazil 1980
42. Sitton, O. C. et al.: Chem. Eng. Progr. *75* (12), 52 (1979)
43. Clément, C., Rebeller, M., Trambouze, P.: Proc. of the 7th Petroleum Congress, Mexico 1967
44. Allen, A. L.: A.I.Ch.E. Symp. Ser. *42*, 115 (1976)
45. Sezzi, F.: Atti del Simposio Europeo Automobilistico, vol. 2, ECC, Bruxelles 1955
46. Scheller, Wm. A.: Proc. of the 8th Nat. Conf. on Wheat Utilization Research, Colorado 1973
47. American Petroleum Institute: Publication No. 4082, 1971
48. Spencer, E. H., Brandberg, A.: Proc. of the Vth Alcohol Fuels Technology Internat. Symp., vol. 2, p. 159, Auckland, New Zealand 1982

49. Jawetz, P.: in Energy from Biomass, Ist ECC Conf., Brighton, U. K. 1980
50. American Petroleum Institute: Publication No. 4261, 1976
51. Ecklund, E. E.: Paper presented to U.S. Senate Agricultural Research and General Legislation Subcommittee, Indianapolis, Ind. 1977
52. Scheller, Wm. A.: Proc. of the Int. Symp. on Alcohol Fuels Technology, Methanol and Ethanol. Wolfsburg, FGR 1977
53. Brinkmann, N. D.: Preprints of the IVth Alcohol Fuels Technology Internat. Symp., vol. 2, p. 339, Guarujà, Brazil 1980
54. Myburgh, I. S., Kerens, G., Van Bergen, E.: Proc. of the Vth Alcohol Fuels Technology Internat. Symp., vol. 2, p. 421, Auckland, New Zealand 1982
55. A.C.R. Process Corp.: Energy Balance. Part of a preliminary Pilot Project Application to U.S. Dept. of Agriculture 1978
56. Pimentel, L. S.: Biotech. Bioeng. *22*, 1749 (1980)
57. Mednick, R. L., Weiss, L. H., Xippolitos, E. G.: Chem. Eng. Progr. *78* (8), 68 (1982)
58. Meisel, S. et al.: Chemtech (2), 86 (1979)
59. Derouane, E. et al.: J. Catal. *53*, 40 (1978)
60. Chang, M., Anderson, A. W., Tsao, G. T.: Preprints of the 184th ACS Meet., Kansas City, Sept. 16, 1982
61. Houben, H.: Zuckerind. *105* (1), 37 (1980)
62. Brooke, D. L.: Sugar y Azucar *72* (12), 33 (1977)
63. Doldi, S.: Chim. Ind. (Milan) (formerly Giornale di Chim. Ind. Appl.) *15*, 593 (1933)
64. Maiorella, B. L.: Hydrocarbon Process. Int. Ed. *61* (8), 95 (1982)

Biotechnology of Thermophilic Bacteria — Growth, Products, and Application

Bernhard Sonnleitner
Department of Biotechnology, Swiss Federal Institute of Technology, 8093 Zürich-Hönggerberg, CH

The potential use of thermophilic bacteria is discussed with regard to academic basic research and with regard to industrial applications. Deficiencies in basic knowledge concerning physiology, kinetics, taxonomy and phenotypic stability are shown to need further investigation. A comparison of culture conditions in natural habitats with artificial conditions typical for industrial cultivations reveals both the potential and the limits of thermophiles. Kinetics and metabolic regulation are discussed as a basis of modern biotechnology for the production of biomass, volatile products, and thermostable enzymes, as well as for immobilized systems and the application in sludge treatment. The engineering problems arising at elevated culture temperatures with materials, instrumentation, reactor transfer capacities, cultivation systems and genetic stability are shown from a practical point of view. Future trends are centered around technical improvements and the solutions of engineering problems, further investigations concerning metabolic control, especially limitations and inhibitions, and the possible tasks of genetic engineering to produce stable constructions and to transfer and stably express selected genes in nonthermophiles.

1 Introduction

Thermophilic microorganisms are known for a long time, especially the enhanced heat stability of their spores. They easily withstand pasteurization and necessitate sterilization procedures to allow aseptic processes. When organic material is treated even at elevated temperatures infections caused by thermophiles and sometimes serious disturbance of the processes may occur (e.g., production of sugar). But during the last ten to fifteen years, thermophilic organisms have been regarded less and less as pests, and more and more as potentially beneficial organisms. Based on the first quantitative data on the kinetics of a number of thermophilic organisms, but especially because of the high temperature gradients and the expected acceleration of all reactions (according to the 'golden rule' that an increase of temperature by 10 °C would double the reaction rate), it was assumed that the technical applications of thermophilic, and even of extreme thermophilic, bacteria could be highly advantageous, as compared with classic bioprocesses which require temperatures ranging between room temperature and/or about 37 °C. It has also been expected that the rapid growth of thermophilic bacteria would result in greatly reduced cultivation times and volumes.

The production of extremely thermostable biocatalysts should be possible since in several experiments the growth of extremely thermophilic bacteria has been shown to be stable at temperatures quite close to the boiling point in both the natural environment [1–3] and under laboratory conditions (up to 105 °C [4]). Even higher temperatures are thought to allow life and growth if only the medium can exist in a liquid state, as it is the case under high pressure conditions, for example, near deepsea hydrothermal vents, where water boils at 460 °C [5]. Growth at such extreme temperatures implies both an intact structure and function of all the constituents of a cell, such as proteins, membranes, and nucleic acids. Moreover, increased resistance to chemical agents seems to be correlated with increased thermostability in many cases [6,7].

Thermophilic processes could also be considered quite useful for technical and environmental purposes. Cooling problems would no longer exist because of the high temperature gradients, although heating problems might arise if improperly sized or improperly designed reactors are used. Volatile products (e.g. ethanol, acetic acid) could be recovered with less effort, i.e., with lower additional energy input during separation and/or purification. Moreover, possible product inhibition can be easily reduced or even eliminated if the products are withdrawn directly from the culture. The risks of contamination would be greatly reduced, however, as the experiences in thermophilic laboratories show, this expectation is wrong. And last, but not least, special precautions during cultivation, downstream processing, and treatment of residues would not be necessary since organisms that cannot grow below 40 °C are generally not regarded as pathogenic. Not any case of human hazard has been reported so far. Detailed lists of the advantages of using extremely thermophilic bacteria and/or their enzymes have been given [8-14]. The following summary lists those that have been shown realistic, at least in small scale.

- General advantages are: Reduced viscosity of media, which increases the quality of mixing and allows easier liquid-solid separation. A higher solubility of most reactants and accelerated diffusion, which may allow to operate with higher concentrations of reactants. Since many thermophiles are inhibited by 'higher' concentrations of substrates or products this possibility must be carefully checked in every single case.
- Productivity would be increased as the reaction rates (of organisms and enzymes) increase. The increase of rates, however, is in no case as high as expected earlier. Specific growth rates of extreme thermophiles are a factor of only something like 1 to less than 10 higher than those of mesophiles.
- Mass cultivation of thermophilic bacteria would be cheaper than that of mesophilic bacteria due to reduced investment required for heat exchange equipment. The temperature gradients would be sufficiently high to
- accomplish control of reactions by cooling only. This is true when only temperature control is concerned. Due to increased reaction rates the control of, for example, pH and oxygen partial pressure must not be neglected.
- Mass cultivation of thermophilic bacteria would be cheaper than that of mesophilic bacteria due to greatly reduced contamination problems. The latter holds true for very rapidly growing thermophiles and for specialized mixed cultures; it does not hold for less rapidly growing bacteria and for organisms with rather complex nutritional requirements.
- Because of the high temperature of operation, thermophilic enzyme reactors would not be prone to contamination. This must also be critically regarded for large-scale reactors. It has been shown in laboratory scale for a few systems. Without additional detrimental environmental conditions the danger of contamination must be taken into account.
- Cost-effectiveness would increase due to longer useful life of organisms and/or enzymes. This may be true for enzymes in general and for immobilized systems in particular. But free organisms that cannot fulfill their (high) maintenance requirements would quickly die out.
- Higher enzyme yields would result due to greater enzyme stability. This is of particular significance for enzyme recovery from the cultures and for enzyme purification:
- Isolation and purification of thermophilic enzymes could be carried out at ambient temperatures because the activity of highly thermostable enzymes would be reduced to negligible low values at low temperatures. However, contaminating

thermophilic proteolytic enzymes can give rise to serious losses in spite of relatively low operation temperatures.

— Thermostable enzymes generally seem to be more resistant to the denaturing effects of detergents and organic solvents. The consequence is the possibility of applying higher concentrations of reactants, the choice of not necessarily highly aquatic reaction systems, or the possibility of efficient cleaning of the systems with organic solvents. Again, the large-scale feasibility remains to be shown.
— Volatilization of products might be valuable for two reasons: as a recovery step or as a means for removing potentially inhibitory products. This could further be enhanced by applying vacuum to the culture system.
— Anaerobic cultures would be easier to operate due to decreased oxygen solubility. As the experiences show at present, the start-up of anaerobic cultures of thermophiles is also dependent on the correct redox potential which is equally easy or difficult to establish under mesophilic conditions. Increased diffusion rates may partly make up for decreased oxygen solubility.
— (Extreme) thermophiles would not be pathogenic.

The more recent studies have revealed that the early expectations for the usefulness of thermophilic bacteria seem to have been somewhat overestimated. Nevertheless, the production of thermostable and/or chemoresistant enzymes, the production of chemicals and fuels, the degradation of polymeric or polluting substances at high rates, and the direct treatment of sewage sludge (hygienization) are promising projects. However, for their realization, a basic knowledge of life at high temperatures is required.

Much work has been done, and must be continued, to understand the molecular basis of thermophily. Today it is clear that the mechanisms of life, metabolic and genetic control are not principally different from life in the mesobiotic or psychrobiotic range. It is worth noting that the idea of technical application of thermophiles was not originally born by biotechnologists. It was the molecular biologists who were interested in 'why is life at high temperatures possible?' and 'how do or must the cellular constituents look like that they can function and be stable at the high temperatures?'. Their answers and findings then showed the potential usefulness of thermophilic systems and their technically interesting properties. The technical exploitation of thermophiles now shows in return the deficiencies in basic knowledge and gives new impulses for scientific academic and applied research. Since the molecular basis of thermophily is not the goal of this article, a list of excellent reviews is given in Sect. 3.1. From an evolutionary point of view, thermophily is quite interesting. It is assumed that thermophiles evolved from mesophiles and not vice versa. Some of the extreme thermophiles, such as *Sulfolobus*, *Thermoplasma*, *Thermoproteales* and *Methanobacteria*, are no longer classified as 'classic' eubacteria. They are nowadays regarded as members of the archaebacteria [15-17].

The diversity of microorganisms appears to decline with increasing growth temperatures. Eukaryotes have never been shown to grow above 62 °C [18], and photosynthetic prokaryotes have never been observed above 74 °C [19]. The occurrence of *Bacilli* in habitats with even higher temperatures has been known for a long time [20]. However, in the last two decades, a number of non-sporeformers have been isolated and more or less classified (aerobic and anaerobic neutrophiles and acidophiles).

It turns out that they represent an abundant population of aquatic habitats [13,21–37]. Also within the last two decades, or — possibly due to technical problems — even later, a diversity of thermophilic anaerobes have been isolated. Most of them will probably find large scale application in the future because of their important metabolic characteristics: e.g., cellulose breakdown by *Clostridia* [12,14,38–43], production of alcohols by *Thermoanaerobacter*, *Thermobacteroides* and *Clostridia* [11,14,40,44–50] or the production of methane by *Methanobacterium thermoautotrophicum* and other thermophilic methanogens [14,37,51–60]. Among the facultative aerobes, only the *Thiobacillus* species are currently of potential technical use [25,61–64]. As interest in thermophilic applications increases, screening programs will continue to enlarge the number of potentially useful thermophilic organisms available.

As will be shown later, certain limits, caused by the cultivation equipment and/or by the biological properties of the organism used may make it difficult to use thermophiles in the same direct applications of biotechnology as has been presently established using mesophiles. 'Advanced biotechnology' has to be developed and improved. In many cases, e.g., in the production of enzymes, the transfer of the desired genes to well known mesophiles would result in technically and economically feasible processes. Genetic engineering is currently beginning to deal with exactly these possibilities and to solve them. The transfer of genes coding for thermophilic enzymes into mesophiles has been shown possible and the expression of the still thermophilic gene products in or by the mesophile is active. However, the stability of these constructions and their technical usability have to be investigated in the future. It may be much more convenient and economical to work with some few 'standard organisms' that are physiologically, genetically, and kinetically well known. Optimization of media, culture conditions, and culture systems for such organisms would by far not afford the same amount of labour and time than it is necessary with (completely) unknown new genera or species.

2 Definition of Thermophily and Thermophiles

2.1 Definition of Thermophily

The tolerance to or the need for high temperatures is a criterion that does not usually fit satisfactorily into most patterns of taxonomy. However, the generally accepted categorization of thermophiles into thermotolerant, thermophilic and extreme thermophilic species and strains is based rather arbitrarily on chosen cardinal temperatures, as shown in Table 1. In this context it must be emphasized that thermophiles require high temperatures for growth; they are not active in the mesobiotic range. The same holds for the thermophilicity of enzymes: Because of their constitutional and structural properties they can catalyze reactions at high temperatures but they cannot at low temperatures, or, more exactly, their catalytic activity decreases to negligible low values.

Heinen named the extremely thermophilic bacteria *Bacillus caldolyticus*, *B. caldotenax*, and *B. caldovelox* 'caldoactive' [65]. Aragno [21] called all organisms 'thermophilic' bacteria whenever the optimal growth temperature was in the thermobiotic

Table 1. Definition of thermophily by means of cardinal temperatures. An extension of data compiled by Ljungdahl [7]

Technical term	Cardinal temperatures [°C]			Ref.
	minimal	optimal	maximal	
Thermotolerant	<30			[357]
Facultative thermophilic		<45	>45	[21]
Moderate thermophilic			<70 ... 75	[174]
Thermophilic	$\geqq 40$	$\geqq 55$	$\geqq 65$	[357]
	>30	>50	>60	[97]
		$\geqq 45$		[21]
			>55	[20]
			>45 but <70	[170]
Extremely thermophilic or 'caldoactive'	$\geqq 40$	$\geqq 65$	$\geqq 70$(90?)	[97]
			>70	[65]
			near boiling point	[21]
			>70 ... 75	[174]
			$\geqq 65$	[170]

range (over 45 °C). 'Facultative thermophiles' have their minimal temperature in the mesobiotic range (20—45 °C) or even in the psychrobiotic range (below 20 °C). This rather inflexible method of categorizing thermophilic bacteria may lead to formal problems such as in the case of *B. caldotenax*, which has been isolated at 86 °C [66]. It could be grown under laboratory conditions up to 85 °C on a pyruvate medium, but only up to 72 °C on glucose media and on other kinds of media [66,67]. The categorization of this strain to a certain group clearly depends upon the media used. The same holds true for *B. caldolyticus*, which has been isolated from a 84 °C spring [65] and reported to be cultivated at 105 °C [4].

2.2 Taxonomy

According to the latest edition of Bergey's Manual of Determinative Bacteriology [68], only two obligate thermophilic species can be distinguished among the neutrophilic *Bacilli*: *B. stearothermophilus* and *B. coagulans*. However, some strains of this latter species have been found to fail to grow at 65 °C. There is one species under the acidophilic *Bacilli*: *B. acidocaldarius*. The caldoactive *Bacilli* isolated and described by Heinen [65,66], as well as *B. thermodenitrificans*, and *B. schlegelii* are not considered there but seem to be accepted in the literature as separate species. However, Heinen and Heinen [65] have differentiated between the three isolates and *B. stearothermophilus* on the basis of a) temperature range, b) fatty acid pattern, and c) ultrastructure. Each of these groups can be differentiated further according to a) temperature optima, b) sporulation behavior, and c) ultrastructure. *B. caldotenax* (YT-G) could be grown only at temperatures up to 72 °C to 75 °C on all tested carbon sources except on pyruvate, where growth at 84 °C was reported. These findings, however, have not been reproduced thus far. *B. caldolyticus* and *B. caldotenax* (YT-G) have been reported to grow prototrophically, whereas *B. caldovelox* required methionine (Sharp and Atkinson [69]). In contrast, an absolute requirement of *B. caldotenax* for methionine and biotin has been demonstrated [67] and a maximal temperature of 72 °C has been found on all carbon sources tested, including pyruvate (unpublished results).

As suggested by these examples, a classical taxonomy of thermophilic *Bacilli* appears to be problematic. In Bergey's Manual [68] *B. stearothermophilus* is described as markedly heterogeneous. There are uncertainties about its demarcation. The emphasis to grow at 65 °C results in an exclusion of organisms with a maximal growth temperature between 55 and 60 °C from this species although they cannot be distinguished by any other property. Either the distinguishing criteria have been improperly chosen or the stability of the expression of the 'typical' properties has been so low that the resulting taxonomy is meaningless. Changes in properties that previously had been regarded as typical for a particular organism seem to be at least partly responsible for this rather chaotic situation. However, such findings are rarely published (Zuber, personal communication and [70–73]) due to the lack of reliable means for distinguishing between contamination and 'variations' or mutations. Among genera that contain only a few known strains, such problems are, of course, much less confusing than for the genus *Bacillus*. Presently taxonomic problems of thermophilic bacteria cannot be considered as solved [74].

Taxonomies based on the guanosine and cytosine (GC) content of the DNA or on the denaturation temperature of ribosomes as compiled by Aragno [21] and Wolf and Sharp [69] may work for a species (or a family), however, there are so many interspecies differences that a general definition of thermophily based on either of these taxonomies would not be meaningful. *Thiobacillus* strains even show an inverse correlation: the thermophilic strain TH1 has a lower GC content (48%) than the mesophilic *T. thiooxidans* (51%), strain A2 (69.5%), and other non-iron oxidizing *Thiobacilli* (58—66%) [63]. This has also been found for the genus *Clostridium* [75].

Attempts to develop a practical approach for the classification or characterization of thermophilic Bacilli have usually consisted of grouping similar strains together. Allen [20] classified more than 100 strains according to morphological (cells, sporangia) and biochemical characteristics, which resulted in four groups. Klaushofer and Hollaus [76] recorded 68, mostly biochemical characteristics of nearly 90 thermophilic aerobic spore formers, and they calculated simple matching coefficients (according to Sokal and Michener [77]. On this basis, they constructed a dendrogram and an ordered similarity matrix from which four groups could be distinguished with less than 70% similarity between the groups, some of which could be substructured even further. A similar study exists for isolates of the genus *Thermus* [71]. Wolf and Barker [78] and Walker and Wolf [79] determined three major groups based on the studies of biochemical, physiological, and serological properties. Successful attempts to set up a taxonomy on the basis of RNA polymerase structure have also been undertaken within the group of archaebacteria [80,81].

Taxonomy is, of course, not the domain of biotechnologists, but it is definitely a necessary tool for the identification and recognition of organisms of interest. It is important because of the possible instability of characteristic properties of strains, which has been observed in several cases with *Bacillus stearothermophilus*, *B. caldotenax*, *Thermus aquaticus* and several thermophilic isolates (to be discussed in detail in Sect. 6.5).

Of special interest for biotechnologists is the recognition of a distinct strain on the basis of its growth kinetics and/or production kinetics, as characterized by the values of μ_{max}, Y, completeness of substrate utilization, q_S, q_{O_2}, q_P and of the maintenance coefficient. When possible, kinetics should be determined in dilute defined minimal in 'perfect reactors' [82]. Therefore it is important to rule out the possibility of determining the transport characteristics of the reactor since, as shown in Sect. 4, the (specific) reaction rates of extreme thermophiles can be significantly higher than those of mesophiles. An underestimation of kinetic data can easily lead to a misinterpretation of the capacity of an organism. This has been demonstrated with *Thermus* strains, which have been described as 'relatively slow growing' [83], but by using chemostat technique, it could be shown that *Thermus* grow extremely rapidly [84,72]. The kinetic characterization of thermophiles, including their high maintenance requirements, has thus far been determined for only a few organisms. Further characterizations may prove meaningful for economical and technological reasons.

3 Requirements for Growth at High Temperatures

3.1 Requirements Fulfilled by the Cells

To take part in reactions of growth and product formation at specific temperatures, all of the constituents of cells must be both stable and active. This has been shown in many cases for enzymes, ribosomes, nucleic acids, and membrane systems. Thermophilic enzymes are not much different from mesophilic ones. It is difficult to find out the structural and compositional differences that make an enzyme thermostable since the differences of enzymes from different species, that catalyze the same reaction, have the same order of magnitude as the changes of a protein that render it thermostable. Comparison of proteins from similar organisms is therefore necessary. A relatively high content of hydrophobic and ionic interactions contribute to enhanced thermal stability. On the other hand, the enzymes must be sufficiently flexible at high temperatures to do their duty, but consequently, they are not sufficiently flexible at low temperatures; therefore, their activities are most often negligible at room temperature. Nucleotides in nucleic acids from thermophiles are more frequently modified than those of mesophiles. The purpose is again to achieve higher thermal stability. Membranes must be in a liquid-cristalline state to fulfill their functions. Organisms adapt the fluidity of their membranes to the actual growth temperature by changing the pattern of fatty acids ('homeoviscous adaptation') or by synthesizing lipids that have been regarded as nontypical for bacteria. Since biochemical and/or molecular biological aspects are not discussed in this review, the following section gives a list of excellent reviews concerning this aspect.

The structure and regulation of proteins in and the ecology and the physiology of thermophilic microorganisms have been reviewed by Amelunxen and Murdock [6,85,86], Argos et al. [87], Castenholz [83], Degryse [88], Langworthy [19], Ljungdahl [7,89] and Ljungdahl et al. [40], Neurath [90], Oshima [91,92], Singleton [93] and Singleton and

Amelunxen [94], Tansey and Brock [95], Walker [96], Williams [97], Zuber and Zuber et al. [98–103]. The academically and technically interesting phenomenon of the thermoadaptation of enzymes, as postulated by Haberstich and Zuber [104] and Frank et al. [105], could not be supported by further studies [67] and Zuber (personal communication). Additional information on membranes and lipids of thermophiles has been compiled by Baumann and Simmonds [106], Esser [107], Langworthy and Langworthy et al. [108–112], Oshima and Oshima et al. [113,114], Oshima [91], Poralla et al. [115], Pask-Hughes and Shaw [116] and Pask-Hughes and Williams [117], de Rosa et al. [118,119], Searcy and Whatley [120], Wakayama and Oshima [121], Weerkamp and Heinen [122], and information on genetics and evolution has been compiled by Castenholz [83], Friedman [123], Johnson [124], Lindsay and Creaser [125,126], Mitchell [127] and Stenesh [128].

3.2 Environmental Factors

3.2.1 Nutrients

For actual technical application, organisms must be able to grow on relatively inexpensive, easily available, and abundant substrates and utilize them almost completely. This is the case with chemoheterotrophic, ubiquitous [11,70,129,130] thermophiles such as *Bacilli* and *Clostridia*. These organisms have been shown to grow on a variety of mono-, oligo-, and polysaccharides, including cellulosic materials, on peptides, organic acids and alcohols, aromatics and hydrocarbons, as listed by Zeikus [12]. Other nonsporulating, rod-shaped bacteria have been isolated only from aquatic sources such as *Thermus aquaticus*, *T. thermophilus*, and other isolates classified as *Thermus* [27,29–32,34,96,131–137], *Thermomicrobium* [26], *Thermoanaerobium* [36], *Thermoanaerobacter* [138], *Thermobacteroides* [22]. *Thermus* strains have been reported to grow on a number of substrates, although with a spectrum that is by far not as broad as that of ubiquitous bacteria. These substrates include glucose, starch, and some organic acids, but not C_1-compounds, alkanes, or aromatics [131]. Glucose, other monosaccharides and starch are not utilized for growth by different *Thermus* strains [72,83,136], but they grow on glycerol or glutamic acid [136] and on complex substrates [72,83]. According to the analyses that have been done all aquatic habitats contain very little organic material (carbon source) [4,12,66,139–143].

All the above mentioned strains are reported to be obligate heterotrophs, as thus far no lithotrophic growth has been shown. They are found in ecosystems with algal species (in mats) and also together with other bacteria [70,139,141,144]. The heterotrophs are thought to grow slowly [83] on excretion or lysis products of photosynthetic species [13]. A minor source of organic material is, of course, the nearby nonaquatic surroundings of springs or thermal rivers. On the other hand, the composition of the aquatic sources has been reported to be rather constant over time; this has been shown with respect to thermal stress [3,18,25], and can be assumed with respect to nutritional and oxygen availability. Therefore the ability of spore formation does not seem to be a necessary requirement for the survival of the organisms, which would explain the relative abundance of extremely thermophilic non-sporeformers in aquatic habitats.

It is likely that the strict aerobes such as *Thermus* strains grow under oxygen limitation in their natural habitats, rather than carbon limited, which would explain the low growth rates observed or estimated [1,145,146]. Oxygen solubility is greatly reduced at high temperatures and oxygen transfer is usually purely diffusive. At a depth of a certain number of centimeters, oxygen is usually completely consumed [144], as indicated by the growth of strict anaerobes such as *Methanobacterium thermoautotrophicum*, *Methanothermus fervidus*, *Thermoanaerobium brockii*, *Thermoanaerobacter ethanolicus*, or *Thermobacteroides acetoaethylicus* [13,22,36,55,138].

3.2.2 Growth Factors, Ions and Range of Concentrations

3.2.2.1 Organic Material

The extremely thermophilic bacteria of the genus *Thermus*, which are all aquatic organisms, are sensitive to high concentrations of organic material. Brock reported the onset of inhibition at a concentration of $10\ g\ l^{-1}$ [147]. However, a comparison of shake flask and chemostat studies showed different values for the onset of inhibition. In the case of *T. aquaticus*, by taking the final yield as criterion, a linear increase of final biomass concentration occurred with increasing concentrations of meat extract and peptone up to $42\ g\ l^{-1}$ of complex substrates and a sharp decrease of biomass was found at higher substrate concentrations (at $s_0 = 100\ g\ l^{-1}$, no growth was detectable at all) [148].

T. thermophilus was inhibited at much lower values (detectable at $0.7\ g\ l^{-1}$ organic carbon), as demonstrated [72] using chemostat technique. A release of substrate inhibition occurred after diluting fresh medium with the cell-free filtrates from stationary cultures of *T. thermophilus* [72] and *T. flavus* (unpublished results), with the specific growth rate of shake flask cultures used as criterion. The same experiment suggested inhibition of *T. aquaticus* by an unidentified product [148], as shown in Fig. 1.

T. aquaticus has been described as unable to grow on nutrient broth [29], unlike *Thermus* XI [32] and isolate K2 [31]. However, growth on this medium has been successfully investigated [84] with a concentration of 1% organic material considered optimal [66].

Many thermophilic bacteria have been reported to require growth factors that are usually supplied in the form of yeast extract, tryptone, peptone, or other frequently used complex substrates; however, these growth factors are sometimes not interchangeable with respect to their efficiency. Many thermophilic *Bacillus* strains are auxotrophic for methionine. Sometimes other amino acids and vitamins are required [149]. *Thermus* strains AT61 and AT62 need several amino acids and vitamins simultaneously and grow well in 1—3% peptone. The strains are auxotrophic for glutamic and aspartic acid, which may be replaced by a mixture of aspartic acid, isoleucine and proline, as well as biotin, folic acid, and p-amino benzoic acid [34]. *T. aquaticus* has been described as prototrophic, 'although growth was considerably faster in enriched than synthetic medium' [24], but very seldom has been cultivated without an addition of complex substrates. The defined media normally contain 3 to 7 vitamins [131,150,151], which suggests that the vitamins have some still undefined

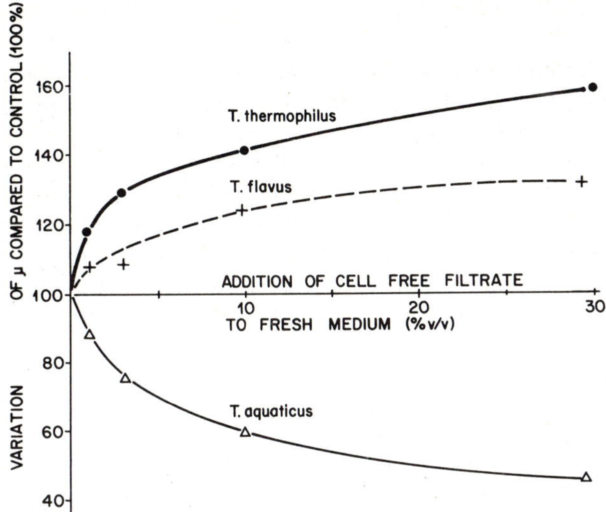

Fig. 1. Evaluation of possible influences of the addition (in % v/v) of filtrates from outgrown cultures to fresh media on specific growth rates (μ) of *Thermus thermophilus*, *T. flavus*, and *T. aquaticus*: (1) dilution of nutrients in the fresh media (2) dilution of (unknown) growth factors in the fresh media (3) addition of (inhibiting) products to the fresh media
For *T. thermophilus* and *T. flavus* the major effect is a release of substrate inhibition due to (1). For *T. aquaticus* (3) is the most probable effect although no inhibitor could be identified. Culture filtrates were prepared from stationary shake flask cultures of the respective strains by successive centrifugation and filtration through membrane filters (0.22 μm pore size). Substrates were not fully depleted since growth occurred upon re-inoculation of the filtrates alone. μ is given in % relative to the control (no addition). Data from [84)] and [72)] and unpublished results

effect. However, the ability to grow without vitamin supplements to the medium has been shown for six *Thermus* strains, using surface culture with the medium solidified with 3% (Difco) agar (no quality mentioned) [131)].

Prototrophy of growth should be shown in submerged cultures because (traces of) impurities in agar preparation cannot be excluded. A white variant of *T. aquaticus*, but not the pigmented strain, was shown to grow prototrophically in chemostat culture [84)]. Using agar surface culture, *T. thermophilus* strain HB8 required only lysine for growth [131)]. These results were not reproduced in chemostat cultures, where *T. thermophilus* grew without lysine, although with low yields [72,152)].

The requirement of *Bacillus caldotenax* (YT-G) for brain heart infusion (BHI) [65,66)] has been shown to be a consequence of auxotrophy for methionine and biotin [153)]. At 'bouillon' concentrations higher than 2%, the yield of *B. caldotenax* decreased. *B. caldolyticus* (YT-P) was less sensitive to higher concentrations of organic material than the strains YT-G and YT-F (*B. caldovelox*) [65)].

The active constituents of yeast extract that enable growth of *Thiobacillus* strains [63,64,154)] are seldom known. In all of the reported cases, yeast extract has been found to inhibit growth (at levels of 0.1%), the optimum being between 0.01 and 0.05%. For strain TH1, only glutathion can replace yeast extract. This strain is inhibited by low concentrations of glucose [63)]. *Thermomicrobium roseum* requires yeast extract and tryptone for growth. A maximal yield is observed when each is provided at 0.5%, but a maximal specific growth rate occurs with 0.3% of each, and inhibition of growth at concentrations of each at more than 1% [26)].

Thermobacteroides acetoaethylicus and *Thermoanaerobacter ethanolicus* both require yeast extract [22,138)]; in the latter case it cannot be replaced by tryptone, casein hydrolysate, beef extract or ashed yeast extract (strains JW200 and JW201). But yeast extract is not reported to function as carbon and energy source or at least its functions are very limited [138)]. *Thermoanaerobacter ethanolicus* grows in media with up to 10 $g\,l^{-1}$ glucose. Mass balances have not yet been evaluated in highly concentrated media [11)]. Yeast extract, and no substitute, is also necessary for growth of *Thermoanaerobium brockii* [36)]. Growth factor requirements of thermophilic *Clostridia* vary. *Clostridium thermocellum* strains (ATCC 27405, LQRI, and NCIB 10682) can grow well on defined media that are supplemented with biotin, pyridoxamine, vitamin B12, and p-amino benzoic acid. Either growth factor of vitamin B12 and p-amino benzoic acid can be replaced by methionine but not both together. Using defined medium, slightly more lactate is produced than in a medium containing yeast extract [155)]. For some strains of *C. thermohydrosulfuricum*, yeast extract or tryptone have been reported to be necessary for growth, but do not serve as a carbon source. The complex substrates do not reduce growth rates up to 2 $g\,l^{-1}$. Lactate, a product, is inhibitory [35)]. *Methanobacterium thermoautotrophicum* is an obligate autotroph, and only cysteine is known to stimulate growth. The addition of a cysteine-sulphide reducing agent decreases the minimal doubling time at optimal growth conditions from 5 to 3 h [37)]. Whether this is a consequence of cysteine addition or of lowering the redox potential is not clear. Yeast extract is not absolutely necessary for growth of *Thermoproteus tenax*. Addition of 0.1 to 0.2 $g\,l^{-1}$ effected a 5 fold increase of the growth rate but 0.5 $g\,l^{-1}$ were found to be inhibitory. A vitamin mixture cannot replace yeast extract for acceleration of growth [156)].

Bacillus acidocaldarius strains have been reported to grow prototrophically [157)], yet yeast extract is normally included in the media [157–160)]. Growth is inhibited by citrate and acetate. However, these nutritional studies were done on agar solidified media using replica plating techniques, and not in liquid culture [157)]. Some other strains isolated from Japanese hot springs require 2 $mg\,l^{-1}$ of biotin for growth in submerged culture [161)].

The facultative chemolithoautotrophic *Bacillus schlegelii* does not require growth factors. Growth is totally inhibited in the presence of 1% glycine and is strongly inhibited when cultures are shaken, regardless of whether autotrophic or heterotrophic conditions are given [162)].

Sulfolobus and *Caldariella* strains grow heterotrophically on yeast extract as energy source. This is not necessary for autotrophic growth [109,163–165)]. Practically all of the thermoacidophiles that are able to leach metals from ores require yeast extract for heterotrophic growth [166)]. *Thermoplasma acidophilum* absolutely requires yeast extract which inhibits growth at concentrations higher than 0.5% (optimal is 0.2%) [167–169)]. Yeast extract cannot be replaced by casamino acids, peptone, or glucose. Of the several hundred compounds tested as substitutes, none support growth, with the exception of a few peptones, ferredoxins and aged glutathione, each of which give only meager responses. Fractionation of yeast extract results in only a minor increase of specific growth promoting activity. These analyses suggest that the growth factor is a polypeptide with a molecular weight of approximately 1000 and containing 8 to 10 amino acids. A protamine or histone-like structure has been suggested but an array of known protamines, histones, or polyamines do not support growth [168,169].

The results obtained with several *Thermus* strains were even less satisfying. *T. thermophilus* could be grown on a defined minimal medium with glucose as the sole carbon source ($2\ g\ l^{-1}$) yielding about $20\ mg\ l^{-1}$ cell dry weight at a dilution rate of $0.5\ h^{-1}$. Yield, but not measurable glucose utilization could be increased by the addition of complex compounds to the medium, such as meat extract, peptones, or yeast extract. However, these substances were by far not completely utilized at all. The search for the actual growth- or yield-stimulating constituents was continued using chemostat pulse technique. None of 20 amino acids, 10 vitamins, several organic acids, fatty acids, amines, polyamines, and sugar compounds tested could replace the complex compounds. Their fractionation was not as extensive as in the case of *Thermoplasma*, but there were no positive results nor were there any indication of which chemical type the assumed growth stimulating substance could be. The search for specific ion requirements that could have been fulfilled by the addition of relatively high amounts of complex compounds was also negative [72].

Thermus aquaticus and several comparable isolates from Icelandic hot springs showed very similar growth behavior. With these strains, extracts from algal bacterial mats and from broken algae were tested for a growth stimulating effect; these attempts had no success nor did the use of filter sterilized spring water to make up the media prove useful [70]. Nutritional requirements for at least some organisms are obviously quite complex, which, as a consequence, results in extremely slow growth rates and/or extremely low yields.

The influences of product concentration on growth and/or product formation have been investigated mainly for anaerobes and sometimes for *Thiobacilli* [63]. The three-fold increase of the rate of ethanol formation in a 1% Solca Floc medium in a coculture of *Clostridium thermocellum* and *C. thermohydrosulfuricum*, compared with the ethanol formation rate of a monoculture of *C. thermocellum*, is due to a release of inhibition by reducing sugars, that accumulate in the monoculture at 10 fold higher concentrations. Coculture of these organisms increases the digestion rate of paper, too [170]. In this process, the ratio of produced ethanol/acetate is increased 8 fold [171]. *Thermoanaerobacter ethanolicus* (wild type) produces the highest amount of ethanol from starch on a 1% medium; however, growth also occurs in a 20% medium [40]. Presently, all thermophilic ethanol producing strains cannot be cultivated without some severe product inhibition at ethanol concentrations higher than 1% ('concentration limitation' [14]), although strains have been known to tolerate up to about 10% ethanol concentrations [9,11,44,170,172].

3.2.2.2. Inorganic Compounds

Usually the tolerance of thermophilic bacteria to inorganic material is typical for non halophilic or slightly halophilic organisms (according to [173]). By generalizing the data available on the salt tolerance of extreme thermophiles, one can say that nonsporulating strains are less resistent to salt than sporeformers (data only on NaCl). *Thermus* strains AT61 and AT62 are totally growth inhibited in media containing 5% NaCl [34]; the same has been observed for *T. aquaticus* [24]. *T. thermophilus* grows poorly in 2% NaCl and is completely inhibited in media with 3% or 5% NaCl [27]; the optimal concentration determined in a yeast extract/polypeptone medium is reported to be 0.2—0.3% NaCl [174]. For *Thermus* Z05, a phosphate concentration of more than 10 mM has been determined to be inhibitory, and CO_2 (resp. HCO_3^-) is assumed to be necessary for growth initiation. This strain also requires glutamic acid as a nitrogen source; it cannot be replaced by NH_4^+ salts. Six other *Thermus* strains tested have also been reported to be inhibited by phosphate at concentrations higher than 10 mM [131]. *Sulfolobus* strains are sensitive to NaCl, with growth being completely inhibited by concentrations of around 1% [175]. Growth of *Methanobacterium thermoautotrophicum* is inhibited by NaCl at concentrations above 100 mM (0.58%) and the specific growth rate is reduced to 70% of μ_{max} at 200 mM. However, CO_2 reduction to methane and growth have been shown to depend on Na^+ with an apparent K_s of about 1 mM [176].

On the other hand, thermophilic *Bacillus* strains grow mostly in media containing 3% NaCl, as shown by Klaushofer and Hollaus [76] for some 80 isolates, all of which are presumably of the type *B. stearothermophilus*, or for *B. schlegelii* [162]. However, growth is inhibited by 5% NaCl. The same relations seem to hold for the anaerobic thermophiles: *Thermobacteroides acetoaethylicus* does not grow in 2% NaCl [22], whereas *Clostridium thermohydrosulfuricum* is not affected by this NaCl concentration; however, its growth is inhibited by 2% NH_4Cl [35]. An exception is *Methanococcus thermolithotrophicus* which requires NaCl for growth (1.3—8.3%); the optimal concentration is near 4% [177].

Besides the usual trace element solutions supplied (e.g. after Castenholz [141], Allen's solution [178], and several similar others), rather unusual trace requirement have been found in a few cases: Traces of Ni [40,179–184] or Se and W [40,185,186] are required to form functioning enzymes. However, this is more typical for anaerobic than for thermophilic organisms [180]. These requirements are normally fulfilled if at least one constituent of the medium is of less than analytical grade quality (compare [153] for Ca requirement of *Bacillus caldotenax*).

So far, thermophilic nonsporeformers, which are typically from aquatic habitats, may be regarded as adapted to a low-concentration environment and thus inhibited by 'higher' concentrations of both organic and inorganic material. However, at higher concentrations, thermophilic sporeformers can still grow. There is no indication known today that the concentration limits for thermophiles are very different from those of their mesophilic counterparts.

Among the thermoacidophiles — eubacteria and archaebacteria — some strains tolerate high concentrations of heavy metal ions. They are of potentially technical interest for microbial leaching at high temperatures. Strains of *Sulfolobus* (*acidocaldarius* and '*Ferrolobus*') have been shown to leach several ores at 60 °C or more, to easily withstand concentrations of 3 $g\,l^{-1}$ Fe^{++}, 2.5 $g\,l^{-1}$ Cu^{++}, and to tolerate up to 2000 ppm Mo-VI [163]. Thermophilic *Thiobacilli* are resistant to $>$1000 ppm Cd^{++} and are tolerant to nearly all heavy metals except for Ag, Hg and oxy-anions of metals [63]. Sulphide, sulphite and tetrathionate are reported to inhibit growth, *Thiobacillus* TH1 by 100 $mg\,l^{-1}$ $Na_2S_2O_3$ or $K_2S_4O_6$ [63] and the neutrophilic *Thermothrix thiopara* by 5 $mg\,l^{-1}$ HS^-, although the HS^- concentration in the natural habitat from which the organism was isolated is 8 ppm. When grown heterotrophically, this organism requires other reduced sulphur sources [23]. Of special interest for industrial application are the findings that *Clostridium thermohydrosulfuricum* is 'relatively tolerant' to high heavy metal ion concentrations, acid and base shocks, as well as to vacuum and O_2 [11].

3.2.2.3 pH

Brock [18,187] has collected data from hot springs all over the world and found a bimodal distribution with respect to the pH values of the springs: Two peaks, one at pH ~3 and a broader one from pH ~7 to ~8.5, with a minimum in between at pH ~5.5. It may be surprising that all thermophilic bacteria known today, and not only those found in an aquatic habitat, fall into the following two groups:

- Thermoacidophiles, which include strains of *Bacillus acidocaldarius*, *Caldariella*, *Desulfurococcus*, *Sulfolobus*, *Thermoplasma*, *Thermoproteus*, *Thiobacillus*, and the
- thermophilic neutrophiles, which include strains of *Bacillus*, *Clostridium*, *Methanobacterium*, *Methanococcus*, *Methanothermus*, *Thermus*, *Thermomicrobium*, *Thermoanaerobium*, *Thermobacteroides*, *Thermoanaerobacter*, and *Thermothrix*.

The first group has pH optima of between 1.5 and 4, the latter between 5.8 and 8.5. True alkalophilic thermophiles have not been isolated so far.

Data concerning pH minima, optima, and maxima in the literature should be critically evaluated. Often the conditions of pH determination are not given, i.e., whether the pH of a medium was set to distinct values prior to sterilization or was measured during the cultivation, or whether the pH was measured using an electrode or indicator paper. Papers may also give incorrect results at higher temperatures. Readings from pH-electrodes are only useful if additional information is given: the temperature of measurement and temperature of calibration of the electrode are the minimum requirements. Several suppliers of calibration buffers give the respective pH values at a given temperature between 0 and 100 °C. Various buffers exhibit different temperature coefficients. Not obeying these simple rules or disregarding the dependence of pH on temperature may easily lead to errors of as much as 1 pH unit.

3.2.2.4 Oxygen; Partial Pressure and Dissolved Gas Concentration

One of the advantages of thermophilic anaerobic processes over mesophilic ones is that they are easier to handle and to control due to reduced oxygen solubility [13]. On the other hand, this is a real problem for thermophilic aerobic cultures. Fig. 2 shows the saturation concentration of O_2 in pure water at 1 atm of air. In the natural habitats, oxygen is transported into the medium only by diffusion. Provided that an aerobic population utilizes this low amount of oxygen, the percentage of total medium allowing growth of strict aerobes is relatively small, as has been shown for a natural ecosystem [144], and provides sufficient oxygen for only very dilute cultures of strict aerobes [18].

At present, very little quantitative data are available on oxygen sensitivity and/or requirements with respect to oxygen partial pressure or dissolved oxygen concen-

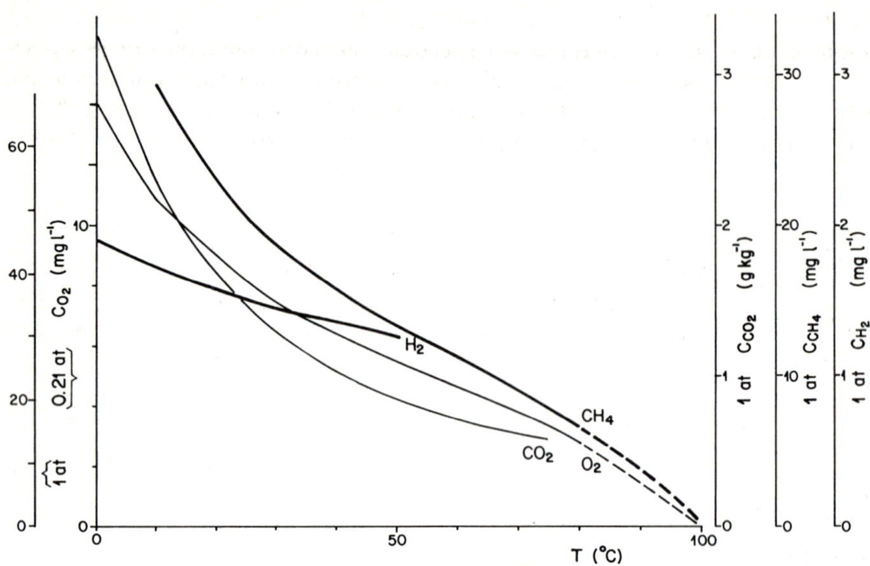

Fig. 2. Temperature dependence of gas solubility in pure water for oxygen, hydrogen, carbon dioxide, and methane at partial pressures of 1 atm, for oxygen also at a partial pressure of 0.21 atm, equivalent to 1 atm of air. Plotted from [381)]

trations. For the nonsporulating isolate 53, which is assigned to the genus *Thermus*, the critical oxygen partial pressure and the optimal pO_2 have been determined to be 6.9% and 84% of air saturation, respectively. The corresponding specific growth rates in batch culture were $\mu = 0.05$ and $0.73\ h^{-1}$ [71)].

The widely used general descriptions found in the literature such as 'strict (an)aerob' [24,37)174)], 'tolerate exposure to air for ... hours' [11,36)], 'vigorous shaking ...' [162,174)] or 'shaking strongly inhibited growth' [162)], where the influence of gas transport remains unclear, too, should be replaced by quantitative measurements.

Data on critical oxygen partial pressure or on the dependence of specific growth rates and/or specific oxygen uptake rate on the dissolved oxygen tension are desirable for reactor and process design, possibly at earlier stages of process evaluation. Yet, their determination is rather problematic (see Sect. 6) but hopefully will be feasible in the near future.

3.2.2.5 Hydrogen; Partial Pressure and Redox Potential

Unfortunately, very few quantitative measurements of redox-potential for cultures of thermophiles are available [51,188–191)]. Indirect information may be drawn from experiments using resazurin, which is generally implemented in media for anaerobic organisms to indicate redox potential below or above +0.042 V at neutral pH [192)]. However, resazurin color is not only dependent on the redox potential of the solution; it is also used as pH indicator and changes from violet at pH 5.8 to orange at pH 3.8. Only a rough estimate of the reduced state of a culture should be drawn from resazurin color. An indication of more than +0.042 V may often be only an explanation for why a culture no longer grows, since the growth and product formation of many strict anaerobes is dependent on redox potentials of less than —0.33 V [192]. The critical redox potential that must not be exceeded is not a constant for a given organism; it varies with inoculum size and richment of the medium [193].

At present most available information must be drawn from data on hydrogen partial pressure. However, this can be troublesome if the organisms are inhibited by hydrogen. Total inhibition of growth of *Thermoanaerobacter ethanolicus* occurs if 75% or more of the gaseous atmosphere consists of hydrogen, the inhibition being detectable from 10% upwards [138]. *Thermoanaerobium brockii* is totally inhibited in the presence of 1 atm hydrogen [14], whereas *Thermobacteroides acetoaethylicus* is not affected by even 2 atm [22] and *Clostridium thermocellum* is not affected by 1 atm of hydrogen gas [14].

4 Kinetics and Regulation

4.1 Maximal Specific Growth Rate

One of the expected advantages of thermophilic bacteria is their high potential for growth, necessarily combined with their high turnover rates of substrates and products. In Table 2, a list of organisms of potential technical interest is shown with data on observed growth rates (or doubling times) at the respective optimal temperatures and pH values and on yield coefficients.

The striking differences among the growth rates of *Thermus* strains determined in different culture systems cannot be easily interpreted. *Thermus* strains, typically nonsporulating, aerobic aquatic thermophiles, have been thought to be rather slow growing [83] and Kandler (personal communication), with one exception: Oshima and Imahori reported a very low doubling time of 18—20 min for *T. thermophilus* [28]. However, most of the μ_{max} determinations have been made so far in shake flasks or in standing flask cultures. Only a small amount of kinetic data has been determined using chemostat technique. As apparent for the *Thermus* strains, the application of chemostat techniques has resulted in quite different data and new aspects of the potential capacities of thermophiles.

T. thermophilus and *T. aquaticus* were cultured in shake flasks with $\mu_{max} \sim 0.2\ h^{-1}$, in controlled batch reactors with μ_{max} not higher than $0.5\ h^{-1}$, and in chemostat, using throughout the same media and culture conditions, with $D_c = 2.7$ and $3.5\ h^{-1}$ respectively [72,84]. Moreover, *T. thermophilus* was also characterized as growing with $\mu_{max} = 0.8\ h^{-1}$ in a controlled batch reactor [194]. In another study [152], the x-D-diagram did not exceed $D = 0.15\ h^{-1}$. Several other details (see Sects. 4.3—4.5) suggested that different strains were used in the different studies; however, the dependence of specific growth rate on the culture system [72,84] remains to be explained, since the differences are significant and cannot be attributed to analytical errors.

In general, one can state that the most rapidly growing bacteria are found among the thermophiles. Thermophiles have the capability to grow quite rapidly (under

Table 2. Kinetic data on maximal specific growth rate (μ_{max}) and/or doubling time (t_d) and cardinal values concerning growth temperature (T) and pH of thermophilic bacteria

Organisms	Media	μ_{max} [h^{-1}]	t_d [min]	T_{min} [°C]	T_{opt} [°C]	T_{max} [°C]	pH_{min}	pH_{opt}	pH_{max}	Yield*) [$g\ g^{-1}$]	Ref.
Bacillus acidocaldarius	Glucose + YE	1.24	33	45	60—65	70	2	3—4	6		157)
	Glucose + YE	0.40			60			3			344)
	Glucose + YE	1.16	36	45	65	75	3	3.3—4	6.2		380)
	Glucose + biotin			40	65	70	2.2	2.5	5		161)
sp. 11-1S	Starch + YE	1.25		45	65	70	2.5	3—4	5		282)
Bacillus caldolyticus	Glucose + BHI				72	80		6—8			65)
	YE	0.23	180			105				~0.17[1])	4)
Bacillus caldotenax	Glucose	2.2		42	65	72				0.4—0.5	67)
	Complex	3									67)
	Glucose + BHI				70						65)
	Pyruvate + BHI				80	82—84					65)
Bacillus caldovelox	Glucose + BHI				60—70	76		6.3—8.5			65)
Bacillus schlegelii	Autotrophic				70			6—7			162)
	Autotrophic	0.23	180	37—45	70	75—80					301)
Bacillus stearothermophilus	Glucose + YE	4.15	10	45	65	70					358)
	Complex	1.96	21	41	55—58	65		7			359)
	Glucose	1.16	36	41	55	65		7			359)
	Glucose			43.3						0.49	222)
	Glucose				59.3					0.30	222)
	Succinate			43.4						0.46	222)
	Succinate				59.3					0.27	222)
NCIB 8924	Glycerol									'maximal'	360)
1503-4	Glucose + glycerol	1.54	27	<42	57	<59				'very low'	361)
	Complex					69					361)
PH24	Phenol + YE	0.83	50	40	55	72		7			362)
	p-Cresol + YE	0.59	70	40	55	72	>6	7			362)
159	Glucose + YE				58					0.29[2])	363)

Bacillus thermocatenulatus	Complex			35	60—70	78					364)
Bacillus thermoglucosidius	Starch	1.44		40	60	72					303)
	+ complex										
Bacillus isolate	Complex				45—65			6—8.5			365)
Caldariella acidophila											
(MT4)	Basal + YE	0.092			87			3			366, 367)
	+ cas. acids										
MT4	Basal + YE	0.11			85						368)
	+ cas. ac.										
MT4	Basal + YE	0.07		~75							118)
MT4	Basal + YE	0.036				~89					118)
MT3	Basal + YE	0.09			75						366)
	+ cas. acids										
Clostridium	Glucose	0.12			60			7.1		0.17	219)
thermoaceticum	+ complex									5—12.3[3])	219)
										0.85[4])	219)
Clostridium	Autotrophic	0.087	480	36	56—60	69—70	<4.7	5.7	>7.5	0.25[5])[a])[c])	369)
thermoautotrophicum										0.5[5])[b])[c])	369)
	Glucose									2.5[5])[c])	369)
	Glycerate	0.35	120	36	56—60	69—70	<4.7	5.7	>7.5		369)
Clostridium thermocellum	Glucose	0.07			60					0.15	370)
	+ cellobiose										
	Cellobiose	0.29								0.09	155)
	Cellobiose + YE	0.42								0.1	155)
	Glucose	0.3			65					0.11	224)
	Cellobiose	0.4			65					0.21	224)
LQRI	Glucose	0.44		40	62	70					14)
	Cellobiose	0.5									14)
	Cellulose	0.15									14)
LQ8	Cellulose + YE	0.096									41)
	Glucose + YE	0.20									41)
N1	Cellulose + YE	0.10									41)
	Glucose + YE	0.33									41)
H	Cellulose + YE	0.10									41)
	Glucose + YE	0.33									41)

Table 2 (continued)

Organisms	Media	μ_{max} [h^{-1}]	t_d [min]	T_{min} [°C]	T_{opt} [°C]	T_{max} [°C]	pH_{min}	pH_{opt}	pH_{max}	Yield*) [$g\ g^{-1}$]	Ref.
mutant S4	Cellulose	0.033—0.11			50—60					0.08—0.20[4])	50)
ATCC 27405	Cellulose + YE	<0.07			60					0.15	370)
Clostridium thermocellum	Cellulose			50		75				0.8[5])	11,170)
Clostridium thermohydrosulfuricum	Glucose + complex	0.55	75	>30	65	<80					13)
39E	Glucose	0.55		40	65	75					14)
39E	Cellobiose	0.5									14)
	Glucose + YE	0.55	75	40	68	78	5	6.9—7.5	9	0.14	35)
				40	69	78	5.5	6.9—7.5	9.3	0.5—1.5[5])	170)
	Glucose	0.46—0.59	70—90	36—39	67—70	75—78	5.5	6.9—7.5	9.3	1.0[5])	11)
	Glucose			53	68—70	75—81	4.8	6.5—7.2	8.5		197)
	Glucose	0.7			65					0.09	224)
	Cellobiose	0.5			65					0.10	224)
Clostridium thermosaccharolyticum	Glucose/starch				55—60	62—68					11)
Desulfotomaculum nigrificans ssp. *salinus*	YE + 1%NaCl			30—40	55—60	70					371)
Desulfotomaculum nigrificans JW400	Lactate + sulfate					68					170)
Desulfovibrio thermophilus	$S_2O_3^{=}$, $SO_3^{=}$			40—45	65	85					33)
Hydrogen bacterium TH1	Autotrophic	0.68		<35	52	55	5	7	>8		347)
Methanococcus thermolithotrophicus	Autotrophic	0.76	55	30	65	70	6	7	8		177)
Methanobacterium thermoautotrophicum	Autotrophic	0.14—0.23	180—300	40	65—70	75	6	7.2—7.6	8.8		37)
strain Marburg	Autotrophic	0.43	96								200)
strain Marburg	Autotrophic	0.28	150								372)
strain Marburg	Autotrophic	0.69			65			7		1.6[6])[c])	54)
strain ΔH	Autotrophic	0.32	132								200)
strain YTB	Autotrophic	0.28	150		65	<75					13)

Methanosarcina TM1	Methanol	0.069—0.087	480—600	35	50—55	<60				4.5^{6})^{c})	373)
	Acetate	0.058	720	35	50—55	<60				1.8^{6})^{c})	373)
	Methanol + acetate	0.14	300	35	50—55	<60				4.5^{6})^{c})	373)
Methanothermus fervidus	Autotrophic	0.24	170	65	83	97		6.5			55)
Sulfobacillus thermophilus				28	50	58—60	1.9	2.4	3		204)
Sulfolobus acidocaldarius	Complex	0.077—0.11	390—540	55	70—75	85	0.9	2—3	5.8		374)
	Complex	0.088	470	45	75	83	1	3	5.5		366)
	Complex	0.092	450	63	87	89	1	3	5.5		366)
'batch'	Natural habitat		'few hours'		63—92			1.5—2.3			146)
'continuous'	Natural habitat	0.023—0.069	600—1800		63—92			1.5—2.3			146)
Sulfolobus brierleyi	Autorophic			50	87	87		3.5—5			81)
Sulfolobus solfataricus	Autotrophic			45	70	75		1.5—2			81)
Sulfolobus TA1	Basal + YE	0.046	900	55	70	>80	1—2	2—3	5		375)
	Basal + YE	0.07		50	65—75	85		3			376)
Thermoanaerobacter ethanolicus	Glucose + YE	0.3		37	69	78	4.4	5.8—8.5	9.8	0.14	138)
	Glucose/starch	0.42—0.46	90—100		69		4.5	5.5—8.5	9.5	1.9^{5})	11)
	Glucose + YE				72			7.5		1.8^{5})	9)
Thermoanaerobium brockii	Glucose + YE									0.067—0.091	225)
	Glucose + YE				72			7.5		1.8^{5})	9)
	Glucose	0.69		40	65	80					14)
	Glucose + YE	0.42	100							0.7^{7})	229)
	Starch + YE	0.14	300							2.65^{7})	229)
	Glucose + complex	0.69	60	>35	65—70	<85	5.5	7.5	9.5		36)
	Glucose + complex	>1.39	<30							~0.8^{1})^{d})	36)
	Glucose + YE	0.69	60	34	65—68	76					11)
Thermobacteroides acetoaethylicus	Complex/glucose	1.39	30	40	65	80	5.5		8.5	0.1	22)
	Glucose	1.38		40	65	80					14)
strain HTB2	Glucose + complex	1.66	25	>30	65	<80					13)
Thermomicrobium roseum	YE + tryptone	0.13	330		70—75	85	6—7.5	8.2—8.5	8.7—9.4		26)

Table 2 (continued)

Organisms	Media	μ_{max} [h^{-1}]	t_d [min]	T_{min} [°C]	T_{opt} [°C]	T_{max} [°C]	pH_{min}	pH_{opt}	pH_{max}	Yield*) [$g\ g^{-1}$]	Ref.
Thermoplasma acidophilum	Complex	0.17	240	45	59	62	0.96	1—2	3.5		167)
	Glucose + YE	0.14—0.20	210—300	40	59	62	0.5	2	4		169)
Thermoproteus tenax	Glucose + YE + S_o	0.41	102	80	88	96	2.5	5	6		156)
Thermothrix thiopara	$S_2O_3^{=}$	0.39	108	55	73	<85		7			23)
	NO_3^- broth	0.30		62	70—73	77					377, 378)
Thermus aquaticus	Complex				70	78—80					66)
	Complex	0.83	50	40	70	79	6	7.5—7.8	9.5		24)
	Complex	0.83	50	40—45	70	79					379)
	Complex	0.35									32)
	Complex	0.53	78		70—71	80.8				'poor'	29)
	Complex	0.7									133)
shake flask, yellow cells	Complex	0.2		40	78	80				0.17[8])	84)
batch reactor, yellow cells	Complex	0.5								0.12[8])	84)
chemostat, mixed culture	Complex	3.5								<0.05[8])	84)
chemostat, white cells	Glucose	1.6		41	70	75				0.4	84)
Thermus caldophilus	Complex				70—75	82		7			137)
Thermus strain DI	Complex	0.63	66		68—72	77.8					29)
Thermus flavus strain 71	Glucose + complex				65					0.13—0.22	132)
	Glucose + YE				70					0.14[2])	363)
strain AT61	Complex	1		40	70—75	81		6.9			34)
strain AT62	Complex	1		40	70—75	81		6—9.5			34)
Thermus Icelandic isolates	Complex	0.83—0.99	42—50		66—75	79—81					29)
Thermus strain K2	Complex	0.83	50	<40	60	80	7.5	7.6—7.8	10		31)
Thermus thermophilus HB8	Complex	2.1	20		65—72	85					28)
	Complex			40	55—60	75					27)

	Complex	2.1—2.3	18—20	47	65—72	85	5.1	7	9.6		174)
	Complex	1.34	31		72—74	80.9					29)
	Complex	0.83	50		75			7			194)
	Glucose									'barely detectable'	131)
shake flask	Complex	0.2								0.06[9])	72)
batch reactor	Complex	0.5			75			6.9		0.06[9])	72)
chemostat	Complex	>2.7		45	75	85	>5	6.9	<9	0.12[8])	72)
chemostat	Glucose	0.15?								0.31	152)
Thermus X1	Complex	0.63		40	69—71	80					380)
	Complex	0.72	58		70—71	77.6					29)
	Complex	0.6									133)
Thermus ZO5	Glucose	0.14	300			85					131)
	Glutamate	0.69	60			85					131)
	Glutamate	0.35	120								150
	Acetate	0.35	120			85					131, 150)
Thiobacillus TH1	Autotrophic				50		1.5	2			64)
	Complex	0.18	225		50						63)

YE: yeast extract;
cas. acids: casamino acids;
complex: if more than one complex substrate is added to the medium or if no defined carbon source is included;
*) if not otherwise specified: $Y_{X/S}$: g biomass (g substrate utilized)$^{-1}$; substrate as in column 'media';
[1]) based on wet weight instead of dry weight;
[2]) effective yield: g biomass (g utilized substrate)$^{-1}$; incomplete substrate utilization observed;
[3]) $Y_{P/X}$: g product (g biomass)$^{-1}$;
[4]) $Y_{P/S}$: g product (g substrate)$^{-1}$;
[5]) $Y_{P/S}$: mol product (mol substrate)$^{-1}$;
[6]) $Y_{X/P}$: g biomass (mol product)$^{-1}$;
[7]) molar ratio of products: ethanol/lactate;
[8]) effective yield: g biomass (g organic carbon utilized)$^{-1}$; incomplete substrate utilization observed;
[9]) economic yield: g biomass (g organic carbon supplied)$^{-1}$; incomplete substrate utilization observed;
[a]) substrate is hydrogen;
[b]) substrate is carbon dioxide;
[c]) product is methane;
[d]) only if cultured in a controlled bioreactor

favorable conditions), even though they have not been observed to do so in their natural habitat (where there may be suboptimal or even less optimal conditions) [1,146,195].

4.2 Yield

4.2.1 $Y_{X/S}$ (cell mass per substrate)

To avoid further confusion, it would be of value and meaningful to define the yield coefficient Y before its values are discussed. As the 'effective' or true yield coefficient (Y_{eff}), the ratio of produced biomass ($x—x_0$) to consumed substrate ($s_0—s$) will be used since values one the absolute biomass production rate (r_X) or substrate utilization rate (r_S) to correctly calculate $Y = r_X/-r_S$ are not available for thermophiles. As an 'economic yield coefficient' (Y_{eco}), the ratio of produced biomass to supplied substrate (s_0) will be used. This differentiation is necessary because incomplete substrate utilization has been observed in a few cases (see Sect. 4.5).

In general, the yields determined for thermophiles do not greatly differ from those measured for mesophiles, provided that the substrates are utilized completely. However, the values, if determined, are a small fraction lower than is expected from comparable mesophiles. This will be considered in Sect. 4.4 and is most likely due to the increased maintenance requirements of thermophiles.

In many cases, the addition of complex substrates as source of growth factors has been reported to be an absolute requirement for growth (see Sect. 3.2.2.1). Only in a few cases complex substances do not function as a carbon or an energy source, as for example, in cultures of *Thermoanaerobacter ethanolicus* [138], *Thiobacillus* TH1 [64], or *Clostridium thermohydrosulfuricum* [35]. However, the yeast extract concentration in the medium has been shown to influence the final yield, but has not been linearily correlated [35]. This allows one to calculate yields correctly. The optimal concentrations of yeast extract and tryptone, without any defined carbon source for growth of *Thermomicrobium roseum* is 0.5% each, if final yield is to be optimized [26]. At 70 °C 0.5 g l^{-1} of brainheart infusion are sufficient to achieve the same yield of *Bacillus caldotenax* (YT-G) from 1 g l^{-1} pyruvate at 70 °C, whereas 1 g l^{-1} of BHI is required at 80 °C [65].

A decrease in yield (final yield) is often used as an argument for inhibition of growth. Although this is good for qualitative decisions, it should always be evaluated in combination with the effect of the concentration of the inhibiting substance(s) on the specific growth rate (for quantitative characterization). This is most easily achieved using chemostat culture, as shown, for example, for *Thermus thermophilus* [72], but is nonetheless practicable in batch, too, as has been shown for other *Thermus* strains [71].

Of special technological interest is the increase of yield by shifting concentrations of regulatory agents as with *Thermoanaerobium brockii*. Supplying 100 mM of acetone to the culture increases not only the yield but also the specific growth rate twofold [14].

Growth yields are also influenced by the products formed during cultivation. Growth of *Clostridium thermoaceticum* is inhibited by high concentrations of acetic acid, but product formation is not inhibited to the same extent. During exponential growth, the expected Y = 0.1 is closely approached, failing to values of as low as 0.03 in the stationary phase [190].

In this context many researchers should be encouraged to quantitatively determine yield rather than use '—', '±', or '+' for the characterization of utilization and effectiveness of substrates.

4.2.2 $Y_{P/S}$ (product formed per substrate)

Product yields have been described at least as rarely as the growth yields of thermophiles. This may be the consequence of the following characteristics:

a) Many 'ordinary' products are more or less volatile and are 'lost' at temperatures above 60 °C via exhaust gas unless special equipment is supplied. Therefore these products are not accounted for in carbon balances which usually give less than 100%.

b) Among products of technical interest that are presently known to be produced by thermophiles are ethanol, acetic acid, and methane, products that are nearly exclusively formed by anaerobes where the contribution of complex substrates is often not clear.

c) (Thermostable) enzyme yields are not characterized on a mass basis (as is Y). The more meaningful activities that will be considered here are not compatible with the usual Y-values.

Ethanol is formed in significant amounts by anaerobic sporeformers as well as by non-sporeformers and also — to a smaller extent — by the facultative (? [68]) aerobic *Bacillus stearothermophilus* [196]. Thermophilic cultures for ethanol formation have been recently reviewed by Zeikus et al. [14]. Most important is *Clostridium thermohydrosulfuricum* since this bacterium utilizes glucose with the highest yield thus far reported for thermophiles to give ~1.9 mol ethanol per mol glucose (strain 39E) [13]. Variations of $Y_{P/S}$ between 1 and 1.5 (mol ethanol mol^{-1} glucose) have been observed with *C. thermohydrosulfuricum* [40]. Therefore, this organism seems to be of potentially practical use, especially because it easily sustains high substrate concentrations as 20% sucrose [197]. Comparable is the yield obtained with *Thermoanaerobacter ethanolicus*, which has been reported to produce 1.8 (mol ethanol mol^{-1} glucose) on 0.5% medium [138,198]. It is also able to grow on media containing as much as 20% starch [40]. However, wild type strains presently have the disadvantage of a pronounced ethanol sensitivity; it can be overcome by the production and selection of tolerant mutants. This is a prerequisite for the use of maximum $Y_{P/S}$ in technical processes and has been started, for example, with *T. ethanolicus* strain JW200: Strain JW200 wt forms a maximum of ethanol in only 1% carbohydrate medium, whereas in the case of strain JW200Fe, the production of ethanol increases up to 14% carbohydrate concentration (no further decrease shown) [40].

Another possibility for increasing the achievable yield is the use of cocultures as demonstrated with *Clostridium thermocellum* and *C. thermohydrosulfuricum*. The former degrades β-1,4-xylans and glucanes, whereas the latter utilizes C_5- and C_6-sugars, as well as dimers, more rapidly, and does not allow a significant accumulation of reducing sugars. The very stable coculture produces ethanol at a rate that is three fold higher and at a fifteen times higher ethanol/acetate ratio than that produced by a *C. thermocellum* monoculture. The coculture reaches essentially the same yield ($Y = 0.58$ g ethanol g^{-1} cellulose) as does the monoculture of *C. thermohydrosulfuricum* in combination with *C. thermocellum* cellulase in a 0.7%

cellulose medium at 100% substrate consumption [14]. The ethanol yield of *C. thermocellum* (mutant strain S-4) on solka-floc SW-100 is dependent on the nature of the buffer used, 0.22 ($g\,g^{-1}$) in $NaHCO_3$ buffer and 0.14 ($g\,g^{-1}$) in phosphate buffer. With increasing temperature (50 to 60 °C), the yield increases from 0.08 to 0.20 ($g\,g^{-1}$), and the maximal specific growth rate from 0.033 to 0.107 h^{-1} [50].

Another product of interest produced by thermophiles in large amounts is acetic acid. However, product recovery from neutral culture broths is the problem, although strains grow and produce very well at neutral pH. *Clostridium thermoaceticum* wild type strains produce about 2 moles of acetate per mol glucose — the theoretical yield is 3: 'homofermentation' [185] — and reach final concentrations ranging from 4 to 15 $g\,l^{-1}$. The highest determined values of 20 $g\,l^{-1}$ (at pH 7) may also be inhibitory both for product formation and for growth. However, an efficiency of glucose conversion to acetate of 90% ($Y_{P/S} = 0.9$, $Y_{X/S} = 0.1$) can be reached with mutant strains (S3 and 1735) even at pH 6 and final product concentrations of 15 $g\,l^{-1}$ [190]. *C. thermoaceticum* (strain 1745), if 'adapted', can grow at pH 4.5 with a doubling time of 36 h and reach a final acetic acid concentration of 4.5 $g\,l^{-1}$. It is expected that further increase of specific growth rate at this low pH can be achieved using chemostat culture. At a pH of 4.5, the criterion of easy product recovery is fulfilled [189,199].

Thermophilic production of methane gas is of great economic importance since the thermophilic methanogens presently known grow much more rapidly than the mesophilic ones. This implies shorter mean residence times and/or smaller volumes needed for biogas formation. Currently only 3 species of thermophilic methanogens are identified: *Methanobacterium thermoautotrophicum* (for comparison of strains, see [200]), *Methanococcus thermolithotrophicus* [177] and *Methanothermus fervidus* [55]. If the rate of methane formation is not limited by both H_2 and CO_2, methane yield is independent of the growth rate ($Y_{P/X} = 0.63\ mol\,g^{-1}$ or $Y_{X/P} = 1.6\ g\,mol^{-1}$) in *Methanobacterium thermoautotrophicum* (strain Marburg), but $Y_{P/X}$ decreases to 0.33 $mol\,g^{-1}$ (or $Y_{X/P}$ increases to 3 $g\,mol^{-1}$) if either H_2 or CO_2 limit growth. This can be explained by the tighter coupling of growth and methanogenesis under substrate limiting conditions; however, the reasons for this are not yet understood [54]. Furthermore, the concentration of NH_4^+ (nitrogen source) triggers methane yield. In limited cultures, the highest productivity is achieved with $Y_{P/X} = 0.47$ to 0.53 $mol\,g^{-1}$ (or: $Y_{X/P} = 1.9$ to 2.2 $g\,mol^{-1}$) versus 0.44 to 0.46 $mol\,g^{-1}$ (or: 2.2 to 2.3 $g\,mol^{-1}$) with excess NH_4^+. Growth rate, but not explicitly methane formation, of *M. thermoautotrophicum* strain ΔH is unaffected by 0.2 M NH_4^+. This and higher concentrations are reported to inhibit methane production in anaerobic digestors [201]. Biomethanation of acetic acid [51] and acid waste water containing acetate and furfural [188] has been successfully carried out at 60 °C using a granular 'consortium' of methanogens and non-methanogens. At a hydraulic retention time of 43 h (dilution rate = 0.023 h^{-1}), 6–8 l of biogas $l^{-1}\,d^{-1}$ containing 55% CH_4 (4.2 l methane $l^{-1}\,d^{-1}$) can be produced from 20 $g\,l^{-1}$ acetic acid (almost equimolar to utilized acetate) at constant pH 6.8. Only part of the granules were found to actively gas at a rate of approximately 1 volume gas per pellet volume per minute [51]. In the previously mentioned waste water, 0.1 to 0.3 l gas $l^{-1}\,h^{-1}$ could be produced at a dilution rate of 0.015 h^{-1}. The authors regard such a process as feasible on a technical scale [188].

Methane production at the high temperatures >65 °C is possible [55] up to 97 °C using *Methanothermus fervidus* or using still unidentified methanogens at even higher temperatures [5]; however, no quantitative data on yields are available at present.

In the future, thermophilic leaching will possibly provide a rapid method of extracting metal from (minor quality) ores [163]. At present, some thermophilic *Thiobacillus* isolates are known that solubilize 4 $g\,l^{-1}$ Fe from 10 $g\,l^{-1}$ pyrite within 10—15 days. However, strain TH1 oxidizes iron at a rate which is only half the rate of the mesophilic *Thiobacillus ferrooxidans* [202], but this thermophile is highly resistant to high concentrations of Cd (>1000 ppm) [63]. No differences between a thermophilic acidophile at 50 °C and *T. ferrooxidans* at 30 °C have been observed with respect to rate and yield during copper leaching from chalcopyrite waste [60]. *Sulfobacillus thermosulfidooxidans* leaches 2 to 2.1 $g\,l^{-1}$ iron from 2% copper-zinc-pyrite ore supplemented with 0.1% glucose or glutamine within 120 h [204]. The above mentioned strains do not effectively leach at temperatures >50 °C. Strains of the genus *Sulfolobus*, which are able to grow at significantly higher temperatures (up to 87 °C), have been shown to leach 51% of the copper from chalcopyrite containing 29% Cu within 60 days, that is, 43% more than what can be extracted only chemically, i.e., without organisms, under these conditions (60 °C, pH 2—3). Mixed cultures of *Sulfolobus acidocaldarius* and *Ferrolobus* leached 38% of the Cu from chalcopyrite containing 0.31% Cu in 161 days at an average rate of 21 $mg\,l^{-1}\,d^{-1}$, but only small amounts of Zn and Ni (1%) and minimal amounts of Pb (2.6 ppm) were extracted biologically [163].

The product yield of enzymes from thermophiles is difficult to compare with yields from mesophilic organisms because of the changed qualities of the enzymes. Enhanced thermal and chemical resistance may be desired much more than equal or even higher yields on a mass basis. For example, *Clostridium thermocellum* (LQRI) cellulase has a higher endoglucanase/exoglucanase activity ratio than *Trichoderma reesei* (QM 9414) cellulase, exhibits total stability at 60 °C, and is not oxygen labile [42]. Cellulase is produced growth-associated by *Clostridium thermocellum* (strain LQ8) with a yield of 125 endoglucanase units and 300 exoglucanase units per g cellulose MN300 degraded in cultivations at 60 °C with 0.7% initial cellulose concentration over 4 days [41].

4.3 Saturation Parameter K_s

Due to a lack of chemically defined media for most thermophiles, caused in turn by the lack of studies on the growth requirements of individual strains, very few kinetic analyses have included saturation parameters. *Bacillus caldotenax* is assumed to have a K_s for glucose of less than 30 $mg\,l^{-1}$ since glucose concentration at half maximal dilution rate was below the limit of detection (30 $mg\,l^{-1}$) [153]. For a white type of *Thermus aquaticus*, K_s for glucose was determined to be 20—30 $mg\,l^{-1}$, whereas yellow strains exhibited much higher K_s values [84]. Growth of *T. thermophilus* was characterized using chemostat culture on complex media. A trial to fit the experimental data with Monod's model suggested apparent K_s values as high as 0.5 g total organic carbon l^{-1} [84]. This supported the assumption that these aquatic organisms adapt their growth rate to the available substrate concentration according to first order kinetics; the concentrations of organic matter in natural habitats is clearly below such K_s values rendering approximately first order substrate uptake kinetics and first order growth kinetics. *Thiobacillus* TH1 has a lower affinity for iron oxidation ($K_s = 7.3$ mM) than *T. ferrooxidans* ($K_s = 0.2$ to 1 mM) [202]. K_M values for ethanol formation from cellulose by a *Clostridium thermocellum* strain were determined to be close to 50 g l^{-1} [50]. The apparant K_s values for both H_2 and CO_2 were measured for *Methanobacterium thermoautotrophicum* (strain Marburg) and were expressed as vol% of the gas in the supplied atmosphere: $K_{s, H_2} = 20\%$ and $K_{s, CO_2} = 11\%$ with concommitant apparent μ_{max} values of 0.49 and 0.61 h^{-1}. From these apparent μ_{max} values, the maximal specific growth rate at 'infinite' concentrations of both H_2 and CO_2 was extrapolated as 0.69 h^{-1} (as presented in Table 2) [54].

4.4 Maintenance of Thermophiles

High maintenance energy is claimed to be a typical characteristic property of extreme thermophiles [12,22,72]. Yet quantitative determination of maintenance requirements need rather long and hard work. Therefore this claim, at present, cannot be regarded as proven, although it seems to be very plausible and reasonable. Quantitative investigations to measure maintenance of thermophiles have been done with 2 *Bacillus*, 2 *Thermus* strains, and with 1 *Methanobacterium*.

At the optimal growth temperature *Bacillus caldotenax* requires 0.68 h^{-1} (exactly: g glucose $\{$g biomass$\}^{-1}$ h^{-1}) and 20 mMol of oxygen g^{-1} h^{-1} for maintenance when grown in defined glucose medium [67]. *B. stearothermophilus* (strain ZN1.2) has a specific maintenance rate of $a = 0.034$ h^{-1} based on dry weight determination or of $a = 0.028$ h^{-1} based on OD measurements on a 'partially defined' [208] medium at 55 °C. This determination is based on a concept of Marr et al. [205], see also [206,207] without reference to the residual substrate concentration [208]. The maintenance coefficient m_S according to Pirt [207] could be derived from the a-value only if the true yield Y_G were known.

Thermus aquaticus (mixed population) exhibits a M_{O_2} value of 16 mMol g^{-1} h^{-1} at the optimal temperature of 75 °C [84]. At the same temperature, *T. thermophilus* has a maintenance requirement for oxygen of $M_{O_2} = 10$ to 15 mMol g^{-1} h^{-1} [72]. Both strains have been cultivated on complex media, and substrates were poorly utilized. No maintenance requirements for carbon sources could be determined. Under quite different conditions, i.e., on defined minimal medium with glucose as sole carbon and energy source, the maintenance coefficients of *T. thermophilus* were determined to be 0.032, 0.050, 0.068 h^{-1} for glucose and 0.63, 0.69, and 0.95 mMol g^{-1} h^{-1} for oxygen at 60, 70, and 78.5 °C, respectively [152]. These values were by more than one order of magnitude lower than the previously mentioned M_{O_2} value [72], although higher values would be expected for growth on a minimal medium: compare the maximal q_{O_2} values of *Bacillus caldotenax*: 35 mMol g^{-1} h^{-1} in complex and 65 mMol g^{-1} h^{-1} in defined minimal medium [67]. However, those measurements were conducted at a dilution rate ranging from $D = 0.02$ to $D = 0.15$ h^{-1} [152] which cannot be representative of the much more rapidly growing *Thermus thermophilus* which has a critical dilution rate $D_c > 2.7$ h^{-1} [72]. Obviously, strains that are kinetically absolutely different, have been used in these two independent studies. This is also suggested by the different substrate utilization characteristics and by the different growth yields reported.

Uncoupling of growth and energy production is one factor among others that increases the value of maintenance; however, uncoupling occurs less frequently than has been generally assumed [209]. A case of maintenance due to uncoupling under non-energy-limiting conditions has been described for *Methanobacterium thermoautotrophicum*. The decrease of product yield, proportional to the energy yield in this case, from 0.625 mol methane g^{-1} cells to 0.3 mol g^{-1} at low energy-substrate concentrations can be explained by a decrease of the maintenance coefficient by a factor higher than the growth rate [54]. Unfortunately, these phenomena have not been quantified nor have the reasons for the differences been elucidated.

The molecular backgrounds of maintenance are not understood at present; however, assumptions that ascribe maintenance to the preservation of correct ionic environment or the (re)synthesis of cell walls or cellular material or mobility seem reasonable. For further details, see [205, 207, 209–217]. From the greatly differing data reported for thermophiles, it may be concluded that high maintenance is a kinetic parameter for clearly distinguishing between thermophiles and nonthermophiles. *Bacillus stearothermophilus* is not an extreme thermophile. The low values [152] of *Thermus thermophilus* are doubtful because they are derived from a growth range that is not representative of the rapidly growing organism and the maintenance coefficient is assumed to be constant over the entire range of dilution rate (= specific growth rate). This assumption has been shown to be incorrect in the case of a *Thiobacillus* [218].

However, more data should be available to answer this question conclusively. On the other hand, high growth and production rates are by far more beneficial than a reduction of maintenance to mesophilic or negligible values would be. It is the productivity that counts. And if primary products have to be produced a high maintenance would be desirable because more product and less biomass can be produced.

4.5 Substrate Limitation of Growth

Complete substrate utilization is a prerequisite for economical processes, i.e., the residual substrate concentration in either a batch or continuous culture must be as low as possible, depending upon the K_s value. Some thermophiles have been found to decelerate and stop growth even at still high substrate concentrations which leads to extremely low economic yield coefficients. Interestingly, such phenomena have been observed and described almost exclusively for aquatic and nonsporeforming thermophiles. *Thermus thermophilus* was found to utilize no more than 45% of supplied complex organic carbon in slow growing shake flask cultures, as well as in chemostat cultures [72]. Upon removal of the stationary cells, a fresh inoculum could be grown on the cell free filtrate, which indicated that sufficient substrate was left behind by the first culture of either *T. thermophilus*, yellow *T. aquaticus*, on glucose [72, 131, 152], but glucose uptake could be quantified only in extremely slow chemostat growth [152]; in other cases, it was 'barely — if any — detectable'. This was commented upon as 'peculiar' [131] or found to be 'within analytical errors' [72]. Other authors found 'practically no growth' [174] or definitely 'no growth' on glucose [29] and poor growth with other single, defined compounds as carbon source [29]. These findings indicate either a high diversity of the species (or genus) or a pronounced variability since all the strains used were said to be derived from the strain HB8 [174,219].

The anaerobe *Thermobacteroides acetoaethylicus* has been found to stop growth although the substrates are not limiting (= not depleted). One reason may be an observed drop of pH to 5.7 which is relatively near the lower limit; however, growth at pH 5.5 has been detected [22]. Such a pH-effect or accumulation of toxic products can be ruled out for *Thermus thermophilus* and *T. flavus* [72]. *Thermothrix thiopara* grows only 2 to 4 generations exponentially and then stops growing even though its substrates are not completely utilized. In this case, it was not elucidated whether toxic products had accumulated [23]. The cessation of growth of *Thermoanaerobium brockii* at high (=nonlimiting) substrate concentrations was explained by the decrease in pH to the lower limit of 5.5 [36].

Unfortunately, too little quantitative data have been accumulated to indicate whether incomplete substrate utilization occurs frequently and constantly among thermophiles and whether this phenomenon is due only to unknown growth factor requirements or is an inherent property of nonsporulating thermophiles. The latter — together with high apparent K_s values — might seem reasonable as a special mechanism adopted by these organisms to ensure survival, that is, by saving the substrate, as a counterpart to the mechanism of spore formation [72].

4.6 Dependence of Kinetic Parameters on Temperature

At the end of the last century, Arrhenius found that the rate constant for chemical reactions changes exponentially with the reciprocal of the reaction temperature:

$$k = k_{\infty} \exp(-E_a/RT)$$

The application of this 'Arrhenius equation' on microbial processes is typical for formal kinetics [82]. Besides this formula, alternative equations may be used to describe the temperature dependence of growth parameters, as for example, in the inclusion of thermal inactivation or the influence of modifiers [220] or, in a simple linear relationship between square root of rate and temperature [221]. This fits experimental data from mesophilic bacteria well.

The most intensive investigations on the temperature dependence of growth parameters have been done using *Bacillus caldotenax* [67]: the Arrhenius plot of specific growth rate ($\ln \mu_{max}$ vs T^{-1}) shows a break at about 45 °C, where the activation energy changes from 196 kJ mol^{-1} at lower temperatures to 76 kJ mol^{-1} in the suboptimal range (up to 60 °C). At the optimal growth temperature of 65 °C, the Arrhenius equation is no longer valid. For this reason, Aragno distinguishes between 'physiological optimal temperature' and 'temperature optimum for growth'. The physiological optimum corresponds to that temperature at which deviation from the Arrhenius relation occurs and is always lower than the growth optimum [21]. In the case of *B. caldotenax*, the two optima differ by about 5 degrees (65 and 60 °C, respectively). In the same temperature range, the Arrhenius relation is valid for the yield coefficients (true yield, corrected for maintenance) and the maintenance coefficients for glucose and oxygen in the case of *B. caldotenax* [67]. However, assuming a concurrence of growth rate and death rate in the optimal and supraoptimal growth range so that $\mu_{observed} = \mu(T) - k_d(T)$ — compare [212] —, the range of validity of Arrhenius's law is extended up to the maximal growth temperature; those values must be regarded as based on viable biomass and not on total biomass measured; see Fig. 3.

The significance of k_d at higher but not at lower temperatures becomes clear from the estimated activation energy E_a for k_d: this is more than twice as high (160 kJ mol^{-1}) as the E_a for μ. Therefore k_d decreases at lower temperatures to non-measurable and negligible, low levels.

The values of M_{O_2} and m_S differ by factors (f) of greater than 10 at the respective optimal temperatures (M_{O_2}: f = 29, $m_{glucose}$: f = 13) when *B. caldotenax* and *Thermus thermophilus* are compared. In any case, the values of *Bacillus caldotenax* are higher than those of *Thermus thermophilus*. A temperature dependence of yield coefficients has been described for *T. thermophilus* [152] and *Bacillus stearothermophilus* [222], that is contrary to that of *B. caldotenax*: The true yield of *B. caldotenax* increases with temperature, while the growth yields of the two other strains decrease (see Fig. 4). At present, this diverging temperature dependence cannot be explained. The behavior of *Thermus thermophilus* is ascribed to the increased proton permeability of the cytoplasmic membrane caused by the increased growth temperature [152].

Depending on quality and quantity of determinations, the dependence of μ values (or the related value of doubling time, which is more frequently presented) on growth

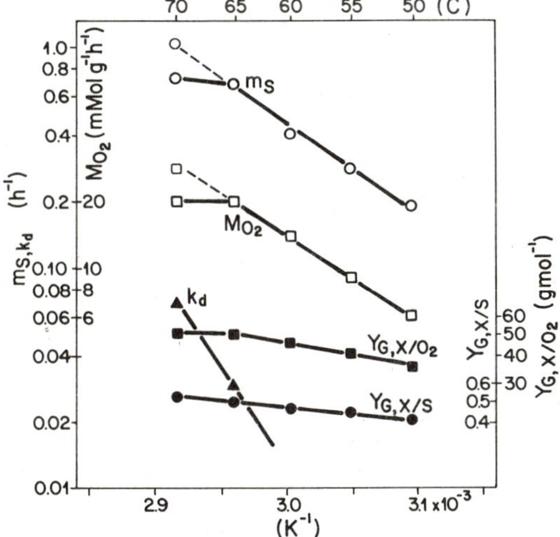

Fig. 3. Temperature dependence of maintenance coefficients (M_{O_2}) and (m_S), true yield coefficients ($Y_{G,X/S}$) and ($Y_{G,X/O_2}$) (i.e., corrected for maintenance), and death rate (k_d) of *Bacillus caldotenax* determined in a chemically defined minimal medium with glucose as sole carbon and energy source. S is glucose. Arrhenius plot. Data from [382] and [67]

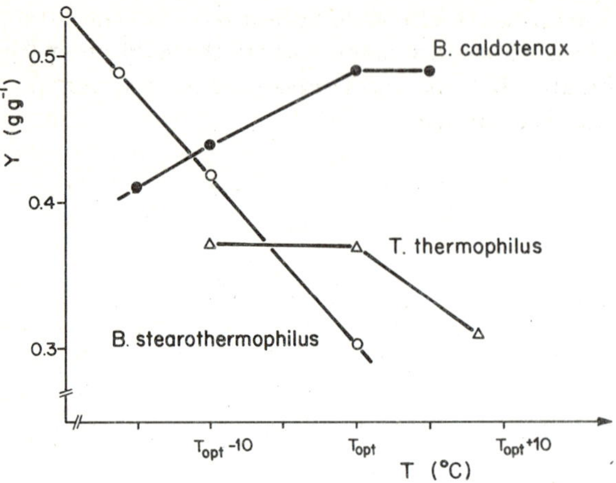

Fig. 4. Temperature dependence of yield coefficients. The plotted values of *Bacillus caldotenax* [67, 382] and of *Thermus thermophilus* [152] are 'true yield coefficients' ($Y_{G,X/glucose}$), i.e., they are corrected for maintenance, and they are determined in chemically defined media using chemostat technique. The values of *Bacillus stearothermophilus* are final yields determined in batch cultures [222]

temperature is more or less of the same type as that described earlier for *B. caldotenax*; for *Thermus aquaticus*, no break in the Arrhenius plot is observed [223]. Yet, two exceptions are presently known: the Arrhenius plots of *Thermoanaerobacter ethanolicus* [138] and of *Clostridium thermohydrosulfuricum* [35] show a plateau at around 60 °C. No explanation for these results is known. However, infections should be ruled out as possible reasons because in both reports, the purity of the cultures is said to have been checked carefully. Unfortunately, no further kinetic parameters of these strains have been investigated for temperature dependence. It would be interesting if product formation and maintenance requirements (and hence effective yields) do not also change over the 5 degree temperature range. This part of research should not be regarded as academic play, as it is of significant importance for process development to know whether temperature changes will affect yield, and if so, in which way (compare Figs. 3 and 4), and to what extent.

5 Products and Application

The following Sections deal with differently investigated subjects. In Sect. 5.1 the knowledge about formation of products studied in laboratory scale is discussed. The respective products are of potential technical interest, however, no larger scale production is reported so far. In Sect. 5.2 the experiences with the production of thermophilic enzymes is reported. It should be kept in mind that thermostable enzymes are presently produced on a large scale using thermotolerant or 'moderately thermophilic' organisms. Thermostability of enzymes from true thermophiles is in most cases higher; these enzymes are not yet produced on a large scale. On the contrary, the thermophilic treatment of sludge (Sect. 5.3.1) has been tested in realistic plants partly operated on a commercial basis. There are many data on the stability of the processes and on gas yields available but the most advantageous thermophilic sludge treatment process is not described decisively.

5.1 Fuels and Chemicals

5.1.1 Ethanol

Ethanol is probably the most promising product from thermophiles at present. Among the organisms forming ethanol in significant amounts are the facultative anaerobe *Bacillus stearothermophilus* [196], and the anaerobes *Clostridium thermohydrosulfuricum*, which is said to be the most potent strain [13,35], *C. thermocellum*, *C. thermosaccharolyticum*, *Thermoanaerobium brockii*, *Thermoanaerobacter ethanolicus*, and *Thermobacteroides acetoaethylicus*. Normal by-products of ethanol production are CO_2, H_2, acetic and lactic acid; however, the ratios of ethanol to by-products can be affected by the choice of culture conditions.

The pure pH-control of a cultivation increases both specific growth rate and specific substrate uptake rate of *C. thermocellum* [35], which has been successfully used in monoculture [38,44] and in coculture together with *C. thermosaccharolyticum* [44] to produce ethanol from cellulosic substrates. The rate of ethanol formation is increased 3 fold when *C. thermocellum* is cocultivated with *C. thermohydrosulfuricum* and the final acetate yield is less than half compared with the monoculture. The enhanced production capacity can be explained by

a) the ability of *C. thermocellum* cellulase to degrade β-1,4-xylans and -glucanes,
b) the capability of *C. thermohydrosulfuricum* to utilize pentoses (and dimers) and
c) a more rapid utilization of glucose and cellobiose by *C. thermohydrosulfuricum*.

As a consequence of lower acetate by-production, the pH does not drop as low as in the monoculture, thereby not inhibiting growth and product formation to the same extent [14,171], and see also [224].

Wild type *C. thermocellum* is 50% inhibited by 0.8% (w/v) of ethanol in the culture broth. However, mutants have been isolated that show the same extent of inhibition, but at about 6% ethanol. In addition, no lactate is produced and the ethanol/acetate ratio is still as high as 5 [44]. Also, more ethanol tolerating strains of *C. thermosaccharolyticum* have been selected and isolated that have produced up to 70% of the theoretical ethanol yield with an ethanol/acetate ratio of 4 [44]. Up to 100% of substrate (=cellulose) utilization can be reached within 120 h at 62 °C using the coculture technique [171]. A different coculture, *C. thermocellum* and *Methanobacterium thermoautotrophicum* is not able to fully degrade cellulose [43]; however, the otherwise long lag time is greatly reduced. The observed interspecies H_2 transport from the cellulolytic to the methanogenic organism is not necessarily required since *C. thermocellum* is not inhibited by H_2. The reduced lag time of the coculture is costly for ethanol production because ethanol yield is decreased to 20% and acetate is increased threefold when compared with *C. thermocellum* monoculture [43]. A coculture of *C. thermocellum* and *Thermoanaerobacter ethanolicus* yields more ethanol and has a higher degree of cellulose utilization than any pure culture [11]. The growth and production rates of *Clostridium thermohydrosulfuricum* are increased in mixed culture with either *Desulfotomaculum nigrificans* or *Methanobacterium thermoautotrophicum* due to removal of lactate and/or hydrogen gas by the cocultured organisms [170].

Thermoanaerobium brockii, a thermophilic lactic acid bacterium, produces high amounts of lactate, but it can be shifted to ethanol production due to the regulatory properties of lactate dehydrogenase (LDH) which is specifically activated by fructose-1,6-diphosphate [225]. The ethanol/lactate ratio partially depends on the yeast extract concentration (~ 1 with $0.5\ g\,l^{-1}$ and ~ 0.5 with $1\ g\,l^{-1}$ yeast extract). *T. brockii* also grows on starch and, compared with glucose as substrate, produces about twice as much ethanol and only half the quantity of lactate, but also twice as much acetate, which is a minor by-product. If an exogenous electron acceptor such as acetone or viable cells of *Methanobacterium thermoautotrophicum* are added to the culture, the yield of ethanol is drastically decreased due to a shift of electron flow over the reversible oxidoreductases [14]. Interestingly, ethanol formation seems to be exponential although growth occurs rather linearly [225]. The extent that the observed cessation of growth under non-substrate limiting conditions [36,225] is due to product inhibition remains to be investigated.

Thermoanaerobacter ethanolicus has been described as producing varying amounts of ethanol (between 1.1 and 1.9 mol per mol glucose), acetate, and lactate; however, the reasons for this are not yet quite clear. The ethanol yield is decreased by higher concentrations of glucose ($>1\%$) and the yield of acids is increased. Further, ethanol yield is affected by alcohol dehydrogenase (ADH) which is highly specific for NADP but inhibited by higher concentrations of NAD [40]. These problems seem to be at least partially overcome by the isolation of ethanol tolerating mutants such as strain JW200Fe [40]. This mutant has not exhibited higher acetate and lactate tolerance. Wild type strains are totally inhibited by 400 mMol of lactate or 450 mMol of acetate. Low amounts of these acids do not enhance ethanol production [138]. Serial transfer to media with higher ethanol concentration allows an 'adaptation' of *T. ethanolicus* so that the organism can still grow in the presence of $90\ g\,l^{-1}$ ethanol [198].

Thermobacteroides acetoaethylicus is the only one of the above-mentioned strains that does not produce lactate as a by-product; however, it produces nearly equal amounts of ethanol and acetate from glucose, independent of the medium supplementation [22]. It is noteworthy to say that among the thermophilic ethanologens, this strain grows by far most rapidly ($\mu_{max} = 1.4$ to $1.7\ h^{-1}$). Substrate level phosphorylation alone is the mechanism of energy conservation which, in combination with high maintenance requirements of this strain [22], opens up the possibility to produce ethanol quite economically, i.e., without wasting substrate for biomass production using stationary cells (compare also [210]). This mechanism remains to be checked for validity for this organism and for the other thermophilic ethanologen bacteria mentioned.

In summarizing the biologically caused features of thermophilic ethanol production, the following observations are important:

— *Clostridium thermohydrosulfuricum* and *Thermoanaerobacter ethanolicus* utilize pentoses (and oligomers, besides hexoses).

- In these two strains, increased H_2 partial pressure does not increase ethanol yield but inhibits growth.
- Theoretically, these two strains form two moles of ethanol from 1 mol glucose (as e.g., *Zymomonas* [226]).
- A direct conversion of cellulosic substrates to ethanol is possible with *Clostridia*, preferentially in coculture.
- Ethanol production with resting cells is possible (however, yeast extract is required in addition to an appropriate carbon source [11]).
- The organisms survive limited exposure to air (oxygen); however, they do not grow and produce in oxygen containing media.
- At least all wild type strains are more sensitive to lower concentrations of ethanol than yeasts (at a level of roughly 1% alcohol). However, tolerant strains have been isolated, which mostly show an improved ethanol/acetate ratio, as well.
- Easy product recovery and improved productivity by applying both high cultivation temperature and vacuum is possible (compare [227]) [11,196].

5.1.2 Acetic Acid

Acetate is formed as a by-product in thermophilic ethanologen cultivations (see Sect. 5.1.1) in amounts depending on the culture conditions but generally not exceeding the ethanol yield. For enhancement of acetate formation, cultural conditions opposite to those needed for ethanol enhancement have to be chosen. However, *Clostridium thermoaceticum* strains are potent producers of acetate and are therefore suitable for industrial scale production: These strains 'homoferment' 1 mol glucose to 3 mol of acetic acid [185]. If mixed substrates are supplied, xylose is utilized first [228]. A prerequisite for economical product recovery is the need to conduct the cultivation at low pH values (as low as 4.5 [199]). High temperatures, of course, support easy recovery. The problem of product inhibition still remains to be solved, as acetic acid is a more effective inhibitor than acetate ion [190].

Acetic acid production is growth associated and then continues during stationary phase. Strains S3 and 1745 reach about 15 $g\,l^{-1}$ of acetate at pH 6 and about 20 $g\,l^{-1}$ at pH 7 with a conversion efficiency of close to 100% (molar ratio of acetate/glucose = 3) [190]. A step-by-step adaptation of *C. thermoaceticum* cultures to both high acetic acid concentrations and low pH (as a consequence of acetic acid and not a different acid) has successfully led to the isolation of an adapted culture (#99—78) that can grow and produce acetic acid at pH 4.5 with a final product concentration of 0.5% (w/v) and still with a 93% conversion efficiency. However, this is achieved at the expense of the growth rate, which at 0.019 h^{-1} is nearly an order of magnitude lower than at neutral pH ($\sim$0.14 h^{-1}) [189,199]. Note that an absolute requirement for successful growth and product formation is a sufficiently low redox potential: at pH 7, at least —360 mV and at pH 4.5, at least —220 mV [189,190,199]; the dependence of redox potential on pH is about 59 mV pH^{-1}. The addition of metals such as iron, molybdenum, tungsten, and selenium increase growth yield (and therefore product yield) [228] because formate dehydrogenase, which contains some of these ions, is otherwise limiting [186].

5.1.3 Lactic Acid

With the exception of *Thermobacteroides acetoaethylicus*, all known anaerobic ethanologenic thermophiles produce lactic acid, mostly as a by-product. Significant amounts of lactate are formed only by *Thermoanaerobium brockii* when the growth medium contains $>1\ g\ l^{-1}$ yeast extract [225]. A representative yield of lactate from glucose is 0.84 mol mol^{-1} [225] or less [229]. In contrast to *Clostridium thermocellum*, *Thermoanaerobium brockii* can grow on ethanol if pH_2 is low, as, e.g., when decreased by *Methanobacterium thermoautotrophicum*, but it then forms acetate rather than lactate. The decrease of pH_2 is necessary because the growth of *Thermoanaerobium brockii* is totally inhibited by 1 atm of hydrogen gas. These findings suggest that lactic acid cannot be produced economically by thermophiles at present.

5.1.4 Methane

The long held assumption that methanogenesis is restricted to temperatures less than 75 °C due to a possible disintegration of inner membrane systems (known from experiments with several strains of *Methanobacterium thermoautotrophicum*) [13] must be revised. This is due to the recent isolation of *Methanothermus fervidus* with a maximal temperature of 97 °C for both growth and methane formation [55] and to the observation that in superheated deepsea water, methane is produced [5]. All thermophilic *Methanobacteria* are strictly autotrophic (they use CO_2 and H_2 and no stimulation of either growth or methane formation occurs upon addition of pyruvate or acetate) [37]. Only cysteine was found stimulatory although not necessary for growth [37]. The amount of acetate converted to methane in thermal volcanic environments is low and its significance uncertain [13]. *Methanothermus* also seems to be an obligate autotroph since formate and acetate do not serve as substrates; however, this extremely oxygen sensitive organism requires yeast extract for growth in an artificial medium [55].

Therefore, thermophilic methane formation is dependent on either geothermal gases or on the production of CO_2 and H_2 by 'precursor organisms' from heterotrophic substrates (compare for example, mixed cultures mentioned in Sect. 5.1.1). Further, the utilization of CO is not likely to be of technological interest since it proceeds at about 1% the rate of CO_2 reduction [230]. On these grounds, a reasonable thermophilic methane production can be based only on the utilization of organic material, preferably wastes, using mixed cultures. This will be discussed later.

5.2 Thermostable Enzymes from Thermophiles

Neither the application nor the molecular basis of thermal stability of such enzymes shall be discussed here, but instead their outstanding properties and the parameters important for process development will be elaborated. Enzymes that are produced world-wide in large quantities primarily include proteases, amylases (including glucoamylase) and glucose isomerase [231–233]. Technical enzymes such as pectinases, cellulases, lipases or oxido-reductases are produced, and presently used in significantly smaller amounts.

Thermophilic organisms, especially sporeformers, have been shown to produce a variety of extracellular enzymes, however, in varying amounts [76]. The mechanisms that control (exo-)enzyme formation in thermophiles are poorly understood so that one can only assume that there is analogy to better understood mesophiles: 'Normally', the maximal synthesis of extracellular enzymes occurs just before sporulation in *Bacillus*, where the causality of the linkage of enzyme production with sporulation is doubtful (at least in special cases) [234–236]. The regulation of exoenzyme formation is assumed to be based on two mechanisms: derepression of catabolite repressed genes and induction of enzymes that are constitutive at a very low level (also shown for *B. stearothermophilus* [94,235,237,238]. In general, complex media have been successfully used for enzyme production, and defined minimal media that support rapid growth are not proper media [239]; see [235] for *B. subtilis* RP1 and [240] for *B. stearothermophilus*. A direct comparison of (mesophilic) industrial production strains and (thermophilic) wild type strains is difficult for at least two reasons:

— Industrial strains have been extensively mutated to obtain the utmost in enzyme yields. It must be assumed that these mutants have lost their 'normal' mechanisms of metabolic control and are no longer comparable to their wild type parent strains.
— Most groups working in this field develop their own enzyme assay systems which have greatly complicated comparisons of quantitative data.

5.2.1 Proteases from Thermophiles

The most famous thermostable protease is formed by *Bacillus thermoproteolyticus* [241–243]. After 12 to 13 h at 53 to 55 °C, the cultivation yields about 4000 caseinolytic units of thermolysin, which differs from *B. subtilis* neutral protease mainly by the higher thermostability (86% remaining activity after 30 h at 70 °C) [244,245]. Unfortunately, this strains is not available from any culture collection. It is thought to be just a variant of *B. stearothermophilus* [246], a theory that is supported by the very similar specificity of thermolysin with a *B. stearothermophilus* protease [247,248] (and a *B. subtilis* neutral protease [249]); however, the maximal temperatures differ by about 10 degrees.

As is also the case with the mesophiles *B. licheniformis*, *B. amyloliquefaciens*, or *B. subtilis*, which yield relatively pure proteases at the end of stationary phase due to a degradation of by-products (enzymes such as amylases) [250,251], the neutral proteases of *B. stearothermophilus* and *B. caldolyticus* are described as maximally excreted in the early stationary phase [252]. Obviously, *B. stearothermophilus* NCIB 8924 produces major amounts of the thermostable (membrane bound) aminopeptidases AP-I, AP-II, and AP-III during growth [240]. In the stationary phase, AP-I remains at a constant level whereas AP-II and AP-III levels decrease during the stationary phase [253]. The formation of protease (and also of amylase) by the *B. stearothermophilus* strain NRRL B-3880 is growth associated [254]. *B. caldotenax* forms a thermolabile protease at the end of the exponential growth phase [240]. Thermitase, a thermostable protease formed by *Thermoactinomyces vulgaris* at 55 °C with an activity maximum between 60 to 80 °C, is produced only in the stationary phase on media containing cornstarch and cornsteep liquor [255–257].

Glucose is known to slow down or totally suppress protease formation. Therefore, media containing lactose [258], starch, grain, fishmeal, enzose-cerelose, casein (hydrolysate), soya fluff, lactosoya or nutrient gelatin are used mostly for industrial scale production [239]. Two-fold higher protease activity is produced by *B. stearothermophilus* NCIB 8924 on Lab-Lemco medium than on CTA-medium [253]. Alkaline proteases from thermophilic enrichments (60 C) have been found to be repressed by free amino acids [258]. *B. caldolyticus* also forms protease in surface culture on glucose [259] and *B. stearothermophilus* on maltose [254]. Caldolysin, a metal chelator-sensitive lytic protease which is stable up to 75 °C between pH 4 and 12, is produced by *Thermus* strain T351 on trypticase-yeast extract medium. This protease seems to be the most thermophilic one thus far described. Its half-life time at 80 °C is 30 h, compared with 1 h at 80 °C for thermolysin and 1.8 h at 73 °C for thermomycolase [135].

Stimulation of protease production has been observed by the addition of rape oil (used as antifoam agent) or acetic acid to *Thermoactinomyces vulgaris* cultures [257]. Calcium ion, an important stabilizer of enzymes [6,135][240][252,254][259–261] must also be present during cultivation, as shown for *Bacillus stearothermophilus* and *B. caldolyticus*; however, different amounts are required. The uptake rate and final amount of Ca^{++} associated with the cells increase with increasing temperature [249]. *B. stearothermophilus* strain NCIB 8924 needs 80—100 mg l^{-1} Ca^{++} and strain NRRL B-3880 only about 20 mg l^{-1} Ca^{++} in a medium providing 3 g l^{-1} maltose as carbon and energy source [254]. These findings indicate that media preparation requires special care. Precaution should be taken to avoid precipitations which may either render some substrates inaccessible to the organisms or may absorb the desired product. Aeration has different effects on enzyme yield in different organisms: 'Relatively low' aeration (i.e. 0.25–0.50 vvm) was found to be optimal for thermitase production [257], whereas an increase of aeration from 0.3 vvm to 1 vvm resulted in a 50% higher protease yield from *B. stearothermophilus*; 0.5 at overpressure increased this activity 16-fold further [240].

The proportion at which *B. stearothermophilus* NCIB 8924 produces AP-I, AP-II, and AP-III is temperature dependent: the highest amounts of the thermostable, Co^{++} requiring AP-I is formed at 55 °C. At 37 °C, the thermolabile AP-II and AP-III constitute the major portion of protease. These two enzymes seem to be released from the cells due to autolysis [253].

Purification of large quantities of enzymes may require special precautions, depending on the degree of purity desired, since autodigestion has been observed increasingly with increasing purity. Only 54 mg of pure neutral protease with a specific activity of 973 units mg^{-1} could be prepared from a 330 l batch of *B. stearothermophilus* [252].

5.2.2 Amylases from Thermophiles

According to the layout of the different enzyme assay systems and the characteristic properties of the enzymes described, it must be assumed that the major activities of the presently known thermophilic amylases are of α-amylase type. More clearly than in the case of proteases, production of amylases by thermophiles is growth associated: On a tryptone-maltose-yeast extract medium, α-amylase can be continuously produced using *B. stearothermophilus* at an optimal dilution rate of 1 h^{-1} with a productivity of 550 units ml^{-1} h^{-1} (dextrinogenic units [262]). For comparison, the critical dilution rate under the applied conditions is about 1.4 h^{-1}; in batch culture, enzyme formation has been determined to be restricted to the exponential growth phase [208]. For *B. stearothermophilus* strain 1503—4, the same observation was reported. The enzyme production yield has been found to correlate to the inverse of specific growth rate which can be controlled by the choice of substrate: the poorer the carbon source for growth (glycerol: $\mu = 0.24\ h^{-1}$, glucose: $\mu = 0.26\ h^{-1}$, and fructose and sucrose: $\mu = 0.42\ h^{-1}$), the higher the enzyme yield (glycerol: 109 U ml^{-1}, glucose: 103 U ml^{-1}, sucrose: 45 U ml^{-1}, fructose: up to 4 U ml^{-1}); compare similar findings for *B. subtilis* [263]. Starch ($\mu = 0.42\ h^{-1}$) and maltose ($\mu = 0.26\ h^{-1}$) do not conform to this relationship, yielding 360 and 220 U ml^{-1}, respectively [237]. From this 'exception', one must conclude that the enzyme formation in this strain is not controlled by endproduct repression. Low molecular weight maltodextrins (maltotriose through to maltohexaose) that are also contained in technical grade maltose increase the differential rate of α-amylase synthesis from 1.2 to 3.0 fold while glucose at concentrations below 10 mM has no effect [238][264].

With strain NRRL B-3880, amylase formation is also growth associated [254]. Additionally, our own experiences with more than 50 amylase producing caldoactive isolates from aerated sewage sludge (manuscript in preparation) suggest that thermophilic (wild type) strains form major portions of enzymes during the growth phase and not in a distinct production (= stationary) phase as has been reported for mesophilic (production) strains [235, 265]. How much this results from insufficient strain development, from unknown regulatory mechanisms, or from unknown nutritional growth- and/or production-requirements of (extreme) thermophiles has yet to be elucidated.

Amylases from thermophiles are mostly (thermo)stabilized by Ca^{++} ions, as has been extensively shown for the *B. stearothermophilus* enzyme [266–268]. Also, *B. caldolyticus* amylase disintegrates into non-active subunits upon Ca^{++} loss [269] but activity can be fully restored by the addition of 0.5 mM Ca^{++} at 70 °C and 1.5 mM Ca^{++} at 80 °C [259]. α-amylase from *B. stearothermophilus* NCIB 8924 (produced at 55 °C) is less susceptible to Ca^{++} loss than the protease formed by the same organism [254]. Without addition of Ca^{++}, the cold labile amylase from *B. stearothermophilus* strain 1503—4 retains no activity at 70 °C, but in 50 mM Ca^{++}, 25% of the 60 °C-activity is retained at 90 °C [262]. The α-amylase from the thermophilic isolate V2 which retains 50% of activity after 1 h at 90 °C, is stabilized by the addition of metal ions in the following sequence: $Ca^{++} \gg Sr^{++} \gg Mg^{++} >$ none $> Mn^{++} > Ba^{++}$ [270,271]. The gene coding for this enzyme has been successfully cloned in a mesophilic *B. subtilis* and is expressed by the host. The transformant is not thermophilic but forms a product that is a little less thermostable than the enzyme from the thermophile; however, the transformant enzyme is significantly more thermostable ($\Delta T \sim 20$ °C) than the *B. subtilis* enzyme [272].

The spectrum of substrates required for amylase formation is interesting. *B. stearothermophilus* strain 1518 does not produce amylase at all in the absence of starch [273]. The enzyme yield in this strain is especially increased by arginine, valine, methionine, Ca^{++}, Mg^{++} and NH_4^+; however, glycerol and vitamins decrease enzyme formation and increase growth. As a technical medium, 2% polypeptone, 0.4% NZ-case, 0.1% starch and 0.5% KNO_3 have been proposed [262]. For *B. stearothermophilus* strain BS1, which forms two isozymes with different thermostabilities, a medium containing potato starch, lactose, and glutamate has been used [274].

Amylase production from thermophilic enrichment is stimulated by free amino acids [258]. *B. caldolyticus* forms amylase on either starch, glucose and brain heart infusion [259] or on amylose plus casein hydrolysate, with the advantage of total protease repression [269]. α-amylase from isolate V2 is produced on starch, yeast extract and polypeptone [270,271]. For amylase production by *B. stearothermophilus*, strain ZN1.2, a medium has been optimized that contains tryptone, maltose, and yeast extract [208]. The enzyme yield from strain ATCC 7954 is increased by the addition of starch, whereas a variation of glucose concentration has practically no effect. Enzyme formation is enhanced by additions of diamalt and KNO_3 [275,276]. α-Amylase production in the mesophile *B. licheniformis* is significantly enhanced by addition of oil seed cakes to the media [277]. Strains 104—1A, 101 and 11—15 of *B. acidocaldarius* produce large clearing zones around cell colonies on starch agar; however, production of the enzyme in submerged culture has been observed to fail if the provided substrate supported growth. Therefore, the enzyme had to be extracted from agar [278,279] using a method similar to the one applied with the *B. caldolyticus* enzyme [65]. The *B. acidocaldarius* enzymes purified this way have been found to be less thermostable than a *B. licheniformis* amylase [280]. However, *B. acidocaldarius* amylase can also be produced in submerged cultures at 60 °C either in standing flasks or with an aeration rate as low as about 0.016 vvm.

Among the substrates tested, only starch (87%), maltose (87%), maltotriose, maltotetraose (50%) and glycogen (100% relative activity) allowed the formation of the enzyme at the end of the exponential and during the stationary phase, reaching a maximum at the beginning of cell autolysis. The strain used in this study (Agnano 101) formed an enzyme that had its maximal activity at 75 °C and pH 3.5 with K_M values of 0.8—0.9 g l^{-1} for starch and glycogen [281]. The α-amylase from the acidophilic *Bacillus* sp. 11—1S was formed only during the stationary phase. It had maximal activity at pH 2 at a temperature of about 70 °C; the K_M for soluble starch was 1.64 g l^{-1}. The effect of either Ca^{++} or EDTA on activity or stability of the amylase was not observed [282].

The experience that strains that are (very) good enzyme producers on surface cultures do not at all or very poorly produce enzyme in submerged culture has also been made by the author with extreme thermophilic isolates from sewage sludge (unpublished) and by H. Sahm (personal communication), too.

A decrease of amylase production during long cultivation has been reported for a mesophilic *B. subtilis* (30 °C, chemostat at dilution rates between 0.05 and 0.15 h^{-1}) after about 5 d [265] and a loss of α-amylase production ability after frequent serial transfers [283]) has also been observed with *B. stearothermophilus* strain 1503—4: instead of 77000 units ml^{-1} expected from earlier studies [284], only about 140 to 200 units ml^{-1} could be detected in cultures of the same (?) organism [262]. In continuous culture of *B. subtilis* NRRL-B3411 a takeover of the culture by mutants that produce less enzyme in relatively short time (60 to 200 h) has been described as occurring on several different media [285]. Similar results — but not to the same extent — were obtained with *Bacillus* strains, which were isolated from sewage sludge, in different batch and in continuous cultures (unpublished results).

Optimization of production media has to be done for each particular wild type or mutant strain [258], and the determination of optimal and/or critical culture conditions still remains to be evaluated for many other thermophiles. Necessary information about pO_2, pCO_2, or pH must be obtained from cultural studies. Formation of amylase from *B. licheniformis* is, for example, immediately decreased if pO_2 falls below 20% of the air saturation value [258]. On the other hand, shaking, and hence most probably 'higher' oxygen transfer to the medium, has been found to be unfavorable for α-amylase formation with *B. licheniformis* strain CUMC 305 [277]. The amylase yield from *B. stearothermophilus* decreases to 40% if only pH is held constant at pH 7.2 [208].

5.2.3 Other Enzymes

Among the great number of enzymes from thermophilic bacteria, only a limited portion seems to be of technical and industrial interest at present; the most important ones have been discussed in the preceeding paragraph.

Extracellular cellulase [186,287] from *Clostridium thermocellum* has some more advantages over mesophilic enzymes than only increased thermostability (stable at 70 °C over 45 min [41]). These include a lack of cellobiase or β-xylosidase activity [41,42], the presence of xylanase and hemicellulase activity [39,286], and a lack of proteolytic activity [41]. The enzyme is not inhibited by moderate concentrations of glucose and cellobiose [12,42,288]. The enzyme production is completely growth associated [41,286], and the cellulase is not inhibited by oxygen and does not require reducing agents [41]. It has a higher affinity to cellohexaose than towards cellopentaose [289]. The ratio of endoglucanase/exoglucanase activity is higher than in the case of *Trichoderma reesei* cellulase. Long chain oligosaccharides are the short term reaction products (15 min) and cellobiose or xylobiose are the long term hydrolysis products (24 h) from avicel, MN300, or xylan, respectively [42]. Mainly cellotriose is detected after 2 h [288]. The enzyme is not inhibited by >10 mM Ca^{++}, Mg^{++} and Mn^{++}, as is *Trichoderma* cellulase [42]. The exoenzyme system formed by a different strain (ATCC 27405) requires Ca^{++}; the enzyme is completely inactivated by EDTA. Furthermore, the enzyme is greatly stimulated by 5 mM dithiothreitol or other reducing agents. A crude *Trichoderma* cellulase (from *T. reesei* QM 9414) acts rapidly on filter paper, less rapidly on avicel, and most slowly on cotton. The *Clostridium* enzyme shows a reverse pattern and saccharifies native cotton more rapidly than the fungal enzyme [287]. In addition, *C. thermocellum* grows more rapidly than *Trichoderma viride* [12,49].

No glucose isomerase from an (extreme) thermophile has been reported so far, which may partly result from the lack of good screening methods [290]. However, *B. coagulans* (NRRL 5649-66, a mutant strain) produces glucose isomerase at 60 °C on xylose as carbon source and *B. stearothermophilus* forms a glucose isomerase on a pepton — corn steep liquor — yeast extract — meat extract medium at 55 °C [290]. The *B. coagulans* enzyme is reported to be most active at 75 °C and pH 7 [291,292], i.e., the enzyme is thermophilic.

Alcohol dehydrogenase (ADH) from thermophiles is an interesting enzyme, not only because of its high thermostability but especially because of its stereospecificity and broad substrate spectrum. ADH from *Thermoanaerobium brockii* has a high stability at high solvent concentrations and is of practical use at 60 °C with up to ≦0.4% v/v alcohol or ketone. Furthermore, it has a broad substrate range including linear and branched alcohols, linear and cyclic ketones [47] with the highest activity for secondary and the lowest for primary alcohols [293]. The same enzyme activity has also been found in *C. thermohydrosulfuricum* [14,293].

In contrast to *C. thermocellum* ADH, which is inhibited by low concentrations of products (NAD and ethanol), is unidirectional (NADH oxidizing) and NAD-limited, but not NADP-limited, *Thermoanaerobium brockii* ADH is not inhibited by low levels of reaction products, is both NAD and NADP linked, and is bidirectional (reversible) [14,46,49]. A coenzyme recycle number (NADP) of 20000 during reduction of 2-butanone by 2-propanol has been reported as the highest value for such reactions. However, the stereospecificity of the enzyme seems to decrease with increasing temperature (9:1 ratio of R and S enantiomers observed for 2-pentanol). Nevertheless, the *T. brockii* enzyme is a si-face specific ADH and therefore a valuable complement of the re-face specific enzyme from yeast. Moreover, the enzyme has been easily immobilized and proven to be stable (2 weeks) on an enzyme electrode used for analytical purposes [47]. The thermal stability of ADH from *Thermoanaerobacter ethanolicus* is even higher. The enzyme can be stored at 70 °C for more than 2 d, is slightly denaturated at 80 °C, and has its maximal activity at 95 °C. Within a very broad substrate spectrum, the highest amount of activity has been found with 2-propanol (16 times higher than ethanol) [40,198]. *T. ethanolicus* ADH is dependent on NADP and is not active with NAD; it does not follow Michaelis-Menten kinetics in the absence of pyruvate [198].

Hydrogenase, as pure enzyme or as part of an enzyme system, may play an important role as specific catalyst in the stereospecific production of chiral chemicals such as hydrogenations of Δ^2-enoates or in the reduction of α,β-unsaturated carbonyls, the regeneration of coenzymes in isolated enzyme systems, or biological H_2 production [294–299]. Hydrogenase from *Clostridium thermoaceticum* has been shown to reduce various artificial electron acceptors; however, its thermal stability is limited, i.e., the enzyme is slowly inactivated at 70 °C, although the highest activity has been measured at 95 °C [300]. A membrane-bound hydrogenase, formed by a *Bacillus* (strain MA 48) under both autotrophic and heterotrophic growth conditions, has an optimal temperature of between 70 and 75 °C [301]. The enzyme from *Methanobacterium thermoautotrophicum* has been shown to be Ni-containing [302] and is proposed as an economical means for the regeneration of NAD(P)H in enzyme catalyzed organic synthesis. The gel-immobilized enzyme shows no loss of activity over two weeks either under H_2 at 25 °C or under Ar at 5 °C [299].

α-glucosidase from *Bacillus thermoglucosidius* is interesting because of its enormous chemo-resistance; its stability in 8 M urea at 25 °C, however, is lost at 60 °C [303]. The enzyme from strain KP 1006 has been shown to be highly resistant to proteolytic enzymes when compared with the mesophilic counterpart from *B. cereus* [304].

Because of their excellent stability and preservability, glucokinase and glucose-6-P dehydrogenase from *Thermus thermophilus* or *Bacillus stearothermophilus* are preferred for analytical and clinical test systems [305].

Oxido-reductases from *Caldariella acidophila* have been studied using intact resting cells, acetonized cells, egg white entrapped or acrylamid entrapped cells for transformations of progesterone, resulting in a conversion efficiency between 4 and 24% (at 85 °C). The product spectrum in this system depends on temperature and oxygen availability. The hydroxylative and oxidative system are both inactive below 70 °C and the hydroxylative system is activated by oxygen [306]. From the same organism, a constitutive β-galactosidase has been found to be stable even over 8 months of storage if cells are entrapped in polyacrylamid gel. Continuous hydrolysis of lactose has been performed with an efficiency of 73% at 70 °C. Under these conditions, the half life is 30 d. Cell entrapment results in a sharp increase of enzyme activity as compared with intact free cells. Acetone treatment causes no or only little change in β-galactosidase activity [10,307,308]. This β-galactosidase has also been successfully entrapped in an albumin-glutaraldehyde crosslinked ultrafiltration membrane [309] and in polyurethane structural foam [310]. No enzymatic loss has been determined during 140 h operations with a conversion degree of 30% at 70 °C and trans-membrane pressures between 0.1 and 4 atm [309,310].

Highly stereospecific asparaginases that are produced from *Thermus* T-351 are important in chemotherapy and for the production of pure, optically active compounds. It is noteworthy that even one of the *Thermus* enzymes is by far more resistant to proteolysis and less cold labile than the corresponding *Escherichia coli* (B) enzyme. Besides higher thermostability [8,311], the specificity of the *Thermus* enzyme for asparagine is exclusive, whereas the *E. coli* enzyme shows activities towards other amino acids, too [312].

Modification methylases from *Thermus thermophilus* or other *Thermus* strains are expected to become useful tools in the preparation of recombinant DNA libraries [313,314].

Among the numerous enzymes from thermophiles characterized, not many seem to be of industrial use at present; however, as pointed out in the previous example, they are likely to become valuable tools (at least) in research.

5.3 Application of Whole Cells and Mixed Populations

This subject is discussed separately for two reasons:

— The products formed by often not well defined populations vary in composition both qualitatively and quantitatively (e.g., biogas).
— The goal of application is not primarily the production of a certain product but rather the degradation of certain substrates or the destruction of certain organisms (e.g., treatment of waste water or hygienization of sludge).

Some applications of defined mixed cultures have been given in preceeding sections. Molecular aspects of methanogenesis will not be discussed in this context, for reviews, see [56,60,315,316].

Because of higher reaction rates and increased biological purification effectiveness, the quantitatively most important application of thermophilic mixed populations is the treatment of wastes and sludges [317]. Furthermore, high temperatures are necessary to drastically decrease the number of potentially pathogenic organisms such as bacteria, viruses, protozoa, trematoda, cestoda, or nematoda [318], especially, if the stabilized sludge is to be used in agriculture and not deposed, which is the case in many countries; in Europe, between 40 and 80% [318], and in the USA, about 20% [319]. Otherwise, a continuous reinfection of animals would occur [320,321]. The biologically produced heat created by thermophilic populations is an economical alternative to pasteurization. The expected advantages of aerobic thermophilic treatment include a relatively low need for external energy (i.e., it is a self heating system), high reaction rates and therefore either low hydraulic retention time or smaller volumes or both, relatively 'good' odour because no organic byproducts are formed in large amounts, and hygienization of the sludges, the degree of which depends — besides on mean residence time — on temperature, pH [318], and on the nature of the substrate.

Aerobic thermophilic treatment may be used as a single step process yielding stabilized sludge and purified water [322] or may be added to conventional waste water treatment processes prior to anaerobic digestion, rendering hygienized, although not fully stabilized, hot sludge. According to our (unpublished) results, the microbial population in aerated sewage sludge at temperatures between 50 °C and 60 °C is quite stable and shows a high degree of diversity with respect to the substrates utilized (proteinaceous materials and carbohydrates: poly-, oligo-, and monomers) and to the possible temperature range of growth; organisms have been detected growing at 4 °C as well as at temperatures greater than 80 °C (in preparation).

In a model study using industrial and agricultural waste effluents that were both high in organic matter, the self heating of the aerated and agitated systems was mainly attributed to the presence of thermotolerant *Bacilli*. Only 7 of the 19 isolated strains could be classified as 'thermophilic' [323]. Batch experiments with molasses as model substrates (26 and 52 l concentrate m^{-3}) showed that during the first 4 to 5 days, temperature increased up to 50—56 °C and then slowly decreased to nearly ambient temperatures. Significant reduction of organic matter (as measured as chemical oxygen demand, biological oxygen demand (BOD_5), and total organic matter) was observed to be directly associated with the heat generation. The yield was calculated to be 14.6 kJ g^{-1} of degraded chemical oxygen demand [324]. Especially in the cases of highly polluted industrial waste water, a thermophilic aerobic process can be useful for a primary pollutant reduction with an efficiency of more than 90%. This is theoretically feasible with retention times of several hours to a few days. The problem that plants must often be operated at only partial capacity can be overcome by dividing the entire (needed) reaction volume into 2 (or, at most, 3) equally sized reactors; a storage of the substrate in the reactor and therefore increasing retention time should be avoided.

Aerobic thermophilic stabilization is recommended for agricultural substrates, for sewage sludges and for the purification of waste waters with high amounts of organic pollution. In these cases this process can economically concur with alternative possibilities of stabilization [325]. Investigations with municipal sludge treated at only 45 °C showed that the liquid-solid separation using either centrifugation or filtration (chamber filter press or screen filter press) works satisfactorily [326].

The aerobic thermophilic process of sludge and/or waste water stabilization is not suitable for the recovery of single cell protein because the major part of the thermophilic population consists of *Bacilli*. These organisms have a relatively high amount of indigestable cell wall materials and thus may enhance the danger of toxin accumulation. If the stabilized biomass should be used as animal feed, the use of pure cultures is definitely necessary [327].

The anaerobic thermophilic digestion of waste has been shown to be quite stable, too: Temperature can be changed abruptly from 55 °C to 60 °C without adverse effects on biogas formation (0.55 l gas l^{-1} at 2% volatile solids feed concentration; 50–60% methane in gas phase), and a change of hydraulic retention time from 9 to 3 d did not decrease gas production, with 60 °C reported to be near optimal. The volatile solid destruction efficiency was determined to be 33, 38, 47 and 50% at the respective retention times of 3, 6, 9 and 12 d at feed concentrations up to 10% volatile solids. At higher feed concentrations, the efficiency was greatly decreased (to 7% at 3 d retention time with 11.7% feed). Most rapid methane formation was determined at a 3 d retention time with 8.2% volatile solids feed (4.5 l methane l^{-1} d^{-1}).

Methane production from cattle waste is reported to be essentially constant over a period of two years. Inhibitory effects at higher feeding rates can be explained as due to a general overabundance of minerals, ammonia (which is toxic above 200 mM) and fatty acids. As a solution to this problem, the combination with wastes that are poor in nitrogen and/or phosphorus (as for example, municipal refuse) is proposed [59], compare also [188, 328]. A higher stability of gas production (i.e., little variation) has been observed in a comparative study using cattle waste at a daily loading rate of 6 g of volatile solids l^{-1} digester and 10 d retention time at 40 °C and at 60 °C, respectively. This is noteworthy especially for the thermophilic digestor which operated over an 8 to 12 month period. Acetate was found to be a quantitatively more important precursor of methane at the higher temperature (with a plus of 5 to 10%). The number of total culturable organisms was 10 fold higher in the thermophilic digestor (1.3×10^{10} ml^{-1}) and the number of methanogens by about 3 fold (9 — 10×10^{8} ml^{-1}), as compared with the 40 °C digestor. However, the specific activity of fatty acid oxidizing bacteria was higher in the mesophilic digestor (by a factor between 1 and 3) [329].

One must state that thermophilic digestion is quite comparable to mesophilic digestion in terms of efficiency and stability but that the expectations of higher reaction rates with thermophiles are not fulfilled, at least in this case of semi-continuous operation. On the other hand, when using solid waste in the form of shredded oil waste and domestic sewage sludge, an increased rate of destruction of waste and conversion to methane has been confirmed at 65 °C, as compared to the same process at 37 °C. After six months of thermophilic operation (30 d mean retention time), the average gas production increased to 0.41 l gas l^{-1} culture d^{-1}, compared with 0.23 l l^{-1} d^{-1} at 37 °C, using identical 2.5% digestor feed at the two different temperatures. Concomitantly, the pool size of volatile acids decreased from 3 to 1.3 g l^{-1}, indicating a different balance of the two populations (acidogens and methanogens) at the higher temperature [52]. The increase of organic acids has been observed as a consequence of either an increase of the loading rate or an increase of temperature. In spite of a high propionate level (15 mM), the anaerobic digestion of a mixture of municipal sewage sludge and slaughter wastes was stable at 60 °C, 5 d retention time and a loading rate of 4.4 g organic dry matter l^{-1} d^{-1}. The resulting biogas had a methane content of 70%.

Anaerobic thermophilic digestion is stable even at loading rates that are much higher than the currently usual values. Long term (4 d) and short-term (2 and 10 h) changes of temperature (ΔT = 3, 10, or 18 degrees) always resulted in a reduction of the gas formation rate, but in every case the process recovered alone without any external manipulation. Therefore temperature changes need not be regarded as serious danger for the thermophilic process [330]. It is noteworthy that there was a satisfactory performance of the digestors at these high acid levels. The concentrations of acetate may be expected to be lower ($\sim$10 mM) in (mixed) anaerobic cultures where K_s values of 15.9 mM at 60 °C and even 5.9 mM at 37 °C have been determined [188]. However, similar high values (>2 g l^{-1}) have been observed at 60 °C and 65 °C independent of dilution rate, and at lower temperatures at retention times as low as 3 d [58]. The increase in conversion can be at least partially attributed to the increased demand for maintenance energy at higher temperatures.

The improved performance of the thermophilic digestor was not only shown by calculation of mass balances but also supported by kinetic measurements of acetic acid as intermediate: the rate of acetate consumption was 3.2 times higher under thermophilic conditions although the level of bacteria in the thermophilic digestor was found to be approximately half of that in the mesophilic stage (based on DNA measurements) [52].

Using newsprint refuse supplemented with primary sewage sludge, maximal productivity was shown to occur at a 5 d retention time at 55 °C and a total solids feed of 50 g l^{-1}. The methane content of the gas was found to increase with decreasing retention time (74% at 3 d and only 55% at 24 d retention time), which can be explained as due to the high solubility of CO_2; consequently, a greater amount of CO_2 leaves the reactor with the liquid phase [331]. Thus, an increase in the dilution rate increases the efficiency of the digestor and upgrades the quality of the biogas formed. However, the dilution rate cannot be arbitrarily incrased. Gas production is reduced by 25% if the retention time is reduced from 5 to 3 d [331]. At retention times less than 12 d at 37 °C or less than 14 d at 60 °C, washout of the organisms was observed, with vapor condensate (20 kg COD m^{-3}, containing up to 400 mM acetic acid, up to 30 mM furfural and up to 40 mM sulfur acids) used as substrate. Nevertheless, in this case, retention times as low as 4 or even less than 3 d could be realized by recycling biomass by either external sedimentation or by filtration [188]. During treatment of beef cattle waste, the optimal retention time with respect to methane formation rate was found to be 3 to 6 d (independent of temperature). Longer retention times gave lower gas formation rates, i.e., a net loss of energy, lower retention times, however, decreased destruction of volatile solids [58].

From a kinetic study using shredded domestic refuse, one could conclude that the retention time should not exceed 10 d at 35 °C or 6 d at 60 °C. With increasing temperatures, two distinct phases become more pronounced: one at low retention times with high rate constants and another one at long retention times and low rate constants which can be explained as 'nothing more than endogenous respiration' [332].

Even higher dilution rates were reported as successful at 55 °C when using downflow stationary fixed film reactors with either draintile clay cylinders or needle punched polyester as inert support. The draintile clay support, with a higher surface to volume ratio (1.41 cm^2 cm^{-3} as compared to 0.86), allowed the higher loading rate, of up to 17.5 kg COD m^{-3} d^{-1} (compared with 13.1). Independent of support and dilution rate, the methane content in biogas was constant between 58 and 61%. A decrease of hydraulic retention time from 2.05 to 0.75 d resulted in an increase in the specific gas formation rate, i.e., an increase of the film methane production rate from 0.7 to 1.7 mol m^{-2} d^{-1}. However, the decrease in COD removal efficiency from 89% to only 84% in the case of punched polyester support was rather low. In the case of draintile clay support, a decrease in retention time from 1.28 to 0.55 d increased the film methane formation rate from 0.6 to 1.5 mol m^{-2} d^{-1} and decreased COD removal from 91 to 86%. The biofilm was evenly distributed over the entire reactor at 55 °C but not at 35 °C. Performance of mesophilic [333] and thermophilic [334] reactors was very similar. It is noteworthy that the thermophilic reactors did not show any sign of process instability even at high loading rates and at (very) short retention times.

At present, it is not yet clear whether reaction rates and conversion efficiency of thermophilic anaerobic waste treatment are really significantly increased at higher temperatures or whether they are practically the same as in mesophilic processes. Presently, the mean hydraulic retention time of sewage sludge in anaerobic mesophilic digestors is between 20 and 30 d. The mean methane content of the formed biogas is between 65 and 70%. Both parameters show a relatively broad distribution. These data must be regarded as representative of the status quo in central Europe [335]. Further work is necessary to answer the question of what specific differences in the type of treated raw materials must also be considered. The easier liquid-solid separation which is expected due to decreased viscosity at higher temperatures [52,336] or a different particle size distribution [327] may not only be a thermal effect, but also, to significant extent, a biological effect; but this remains to be investigated in detail. The effect of temperature on the death rates of potentially pathogens must be explicitly determined in practice, i.e., in the realistic environment of sludge(s). Our results show that the mere analogy to death rate constants, as suggested in the literature [52], does not hold true for samples of thermophilic sewage sludge. Even in sludge treated at 58 °C (during more than 20 h hydraulic retention time), up to about 10^4 potentially pathogenic bacteria have casually been determined (unpublished). At present, this must be attributed to the heterogeneous nature of sewage sludge. Thermal death rate constants determined in a homogeneous, dilute aqueous environment must not be applied directly without criticism to a sludge environment.

In every case, the thermophilic processes are charaterized as stable; however, it may take a considerable amount of time to establish a stable thermophilic population [52,332,334]. One problem should not be underestimated: The heat evolution rate in anaerobic digestors is low (estimated to be 0.0049 W l^{-1} [336] and measured as 0.005 to 0.01 W l^{-1} [337]) and the process might require extra heating. One possibility would be extremely large sized economical digestors that would take optimal advantage of thermophilic self-heating [336]. Another possibility would be the aeration of sludge prior to anaerobic digestion, which may not be of significant expense to the methane in the second stage (U. Keller, personal communication). Autoheating of primary and secondary waste-activated sludge could be achieved by using simple self aspirating aerators and well insulated tanks (reactor volume = 28.4 m^3). During 1.5 years of operation, the slurry temperatures normally exceeded 50 °C and reached a maximum of 65 °C; this was also achieved under cold weather conditions (sludge temperature around 0 °C and air temperature around 20 °C) if substrate with more than 2% total volatile solids was fed. Maximal temperatures were obtained at daily loading rates between 3 and 10 kg biodegradable chemical oxygen demand (approximately 12 to 15 kg total solids m^{-3} d^{-1}) at a 5 d hydraulic retention time and low dissolved oxygen residuals ($\leqq 1$ ppm) [319]. Finally, it must be noted that the increased maintenance demand of thermophiles can be a benefit in this special case: In an aerobic stage, this means more heat and less biomass is produced, in the anaerobic digestor, this means an increased methane production.

A comparison between the mesophilic anaerobic digestor and thermophilic aerobic treatment with respect to their hygienization capacities clearly showed the greater advantages of the thermophilic process, especially when the temperature exceeded 50 °C. A reduction of potentially pathogenic bacteria by thermophilic aerobic treatment — compared with raw sludge — by a factor of 10^1 to 10^2 in the anaerobic digestor is nearly negligible when compared with the 10^5 fold or even higher reduction during 24 h (retention time) at 50—65 °C. In the case of virus analysis, thermophilic aerobic treatment proved also to be better than anaerobic mesophilic treatment in that virus inactivation was complete, except on one occasion. Finally, the aerobic thermophilic digestor was superior to the anaerobic mesophilic one in the reduction of viable parasitic ova, but complete inactivation would still require higher temperatures [338].

In a different study on the recovery of entero-viruses, an aerobic thermophilic digestor (50 °C) was 'without doubt' found to be the most efficient treatment for reducing the titer of infectious virus and by far less time-consuming than anaerobic mesophilic digestion and consolidation. However, both processes gave similar final reduction in virus titer, i.e., no complete inactivation to non-detectable levels [339]. For hygienization, thermophilic processes are necessary. At present, it is not clear whether aerobic or anaerobic or combined (2 stage) processes are more advantageous; however, a combination seems economically favorable for sufficiently large plants since the high temperature can be brought about by aerobic growth and the residual portion of degradable substrates can be converted to biogas.

6 Engineering Problems

Not only do biological reaction rates increase with increasing temperatures, but also chemical reactions, solubilization and diffusion processes and aging of materials, all proceed more rapidly at the high temperatures of thermophilic cultivations. The argument that all materials used for microbiological or biotechnological purposes must sustain sterilization conditions and therefore be of satisfactory quality and stability, regardless of further cultivation temperatures, does not necessarily hold true (except in very short-term experiments), as will be shown in the following paragraphs.

From our experiences and from personal communication with users and suppliers of materials for cultivation in extreme environments, we must conclude that material problems are too often underestimated by microbiologists and biotechnologists. Unfortunately, very little has been published on these topics, presumably because the frustrations mostly lead to doubts about the proper experimental design and the subsequent results. Another underestimation concerns the problems that arise with infections and non-aseptic techniques. Our experiences clearly show that the ubiquitous sporeforming thermophilic species can potentially cause the same danger as mesophilic infections under mesophilic culture conditions. However, we have never observed infections caused by non sporulating thermophiles. These experiences are shared by Zeikus [12] who reports overgrowth of *Thermus aquaticus* cultures by *Bacillus stearothermophilus* in improperly sterilized bioreactors and even growth of heterotrophic clostridial species in autotrophic cultures of *Methanobacterium thermoautotrophicum*. Brock stated in 1969 'that the quality of aseptic techniques of people working with thermophiles deteriorates with time, sometimes with disappointing results' [340]. Consequently, thermophilic culture conditions must not be regarded as a patent solution for improper and non-aseptic working techniques. Hopefully, this chapter can give some hints and will encourage those who have had similar problems to do the same.

6.1 Materials Used for Cultivations

Few of the results discussed above have been obtained from experiments using controlled bioreactors. Most are derived from shake flask, Hungate tube or surface culture experiments. Since glassware can be easily cleaned and reused to a sufficient extent, no problems should arise from using this material. The degradation of glass and the concomitant dissolution of metal ions from glass may be disregarded, at least for organisms comparable to *B. caldolyticus*. This bacterium has been shown to remove 17 $\mu g\ h^{-1}$ of silicon from glass cylinders at 70 °C [341].

Media are often solidified not only for maintenance of organisms but also for biochemical tests of organisms. In this context, one has to consider the purity of the solidifying agent, which is usually agar. Agar is a source of Ca, Mg, and other ions even in purified brands which have a reduced mineral content [342]. Due to organic impurities, media solidified with low-quality agar should not be used for purposes of determining absolute growth requirements. Some organisms may fail to grow on agar-solidified media. For *Methanothermus fervidus*, agar must be replaced by silicate [55].

Serious problems have been reported to occur with steel, especially if several different qualities have been combined in one vessel (e.g., DIN 1.4301, often called 'V2A', and DIN 1.4435, 'V4A'). Cultivations in such reactor vessels may not be reproducible or may totally fail due to the solubilization of metal ions. Replacement of parts of 'minor' quality that may be fixed in the reactor or that may be part of additional equipment, such as electrodes, with parts made of the proper kind of steel are necessary. An additional passivation of the steel surfaces using HNO_3 treatment may bring further improvement [343].

Irreproducible results may also occur due to the utilization of several kinds of plastics for tubing connections or sealing ports. The growth of *Bacillus caldotenax*, *Thermus aquaticus* and *T. thermophilus* has been totally or significantly inhibited by such materials [343].

In any case, biological tests with the organism of interest and the particular material under suspicion should be run to rule out the presence of artifacts and/or to determine the degree of inhibition. These tests can usually be done quite quickly. However, in the case of thermoacidophilic cultivations using media containing high Cl^- concentrations, even highly stable steels may not prove satisfactory (because of pitting). In these cases, the use of different metallic materials (such as Ti) or the covering of steel surfaces with inert materials must be considered.

6.2 Instrumentation

To take the full advantages of controlled bioreactors, various sensors must reliably sustain the cultivation conditions. In particular, the stability and accuracy of electrodes should be checked frequently. Increased diffusion rates of the constituents of the media or electrolytes through membranes or diaphragms at high temperatures are one cause for drifting or noisy signals. Accelerated aging of these parts also causes these problems.

The requirements for maintaining electrodes in a proper state are evidently more difficult under thermophilic than under mesophilic culture conditions, i.e., electrodes 'capable of satisfactory operation over the required range of temperature and pH' are either not available [344)] or need extensive efforts for maintenance. From our experiences, pH- and rH-electrodes with bridged electrolytes are recommended, especially if the culture pH is acidic. The frequent replacement of electrolytes and of the membranes of the oxygen probes has also proved advantageous in that the noise on the pO_2 signal can be reduced (from about 10 to less than 2% of the signal). If a temperature sensor for the automatic compensation of the temperature dependent signal change is built into the probe, one must make sure that the sensor and amplifier are designed to compensate for the actual change of the signal at the respective temperature. For example, at temperatures above about 60 °C, the signal change of the IL-pO_2 probe (Ingold, Zuerich, CH) is no longer linear and therefore must not be corrected using the built-in and linearly working compensation thermistor. The practical use of other ion-sensitive or enzyme electrodes (with thermostable enzymes) in microbial cultivation has not been reported thus far.

It is also noteworthy that the last link of a control system and not only the first one is often directly subject to culture conditions, e.g., the end of acid and/or alkali supply lines of pH control. One has to take care that material is not significantly affected by either acid or base at high temperatures. Stainless steel of quality DIN 1.4435, for example, is visibly leached by 4 N H_2SO_4 at about 70 °C within a few days.

6.3 Heat Transfer

Thermophilic culture temperatures are generally regarded as advantageous with respect to cooling because of the increased temperature driving force (ΔT) which therefore results in less demand for cooling water or the possibility to use warmer than 'typical' 20—23 °C [336)] cooling water. However, extreme thermophilic organisms would possibly need extra energy for heating in small-scale reactors, depending on the scale, geometry (surface to volume ratio), and thermal insulation of the reactors used, as well as the substrates, the culture density, and the specific heat production of the organism. The latter factor remains to be investigated in detail, especially in the case of aerobic thermophiles.

Assuming that there is a linear correlation between specific oxygen uptake rate and specific heat production determined for mesophiles [345)], one must expect sufficient heat production for 'self heating' during growth to compensate for various losses due to insufficient insulation and evaporation, bearing in mind that aerobic thermophiles have higher oxygen uptake rates than mesophiles. No additional heat should be required for thermophilic processes with heat-sterilized media but may be necessary for unsterile (continuous) processes using substrates at ambient temperatures, such as thermophilic treatment of waste water or sewage sludge. Experimental studies using several different substrates, reactor systems, and operation conditions are necessary to answer these questions. A possible economical solution may be the use of very large facilities.

Another fact which may significantly influence the costs of cultivation is the evaporation of water and/or of volatile products. These losses are directly proportional to the gassing rate and may be compensated for by humidification of the inlet gas at the respective temperature. This procedure is not practical especially when large quantities must be treated under sterile conditions.

If an inexpensive means for cooling thermophilic cultures is desired (and is sufficient), one could sparge dry gas to the reactor and remove the respective heat of evaporation from the reactor. Simple and cheaply operating reflux coolers could then remove the heat from the system but recycle water and products, if they are volatile, back to the reactor or into any sampling vessel, if desired. With this simple means of stripping products, either with air or (inert) gas, one has a potential method for lowering the product concentration in the reactor. Ethanol formed by *Bacillus stearothermophilus*, a strain obviously not suitable for industrial production of alcohol, could be enriched in such a distillate up to 76-fold more than in the culture broth. As a consequence, its concentration in the culture vessel was reduced to noninhibiting values (0.07—0.40% v/v) [196]. This is of particular interest for the cultivation of thermophiles on highly concentrated media, since 'concentration limitation' [14] is reported to be the most serious disadvantage of thermophilic product formation at present (compare Sects. 5.1.1 and 5.1.2).

6.4 Mass Transfer

Mass transfer is expected to be facilitated at higher temperatures because of reduced viscosity of aqueous solutions (Fig. 5), reduced surface tension, and increased diffusion. Furthermore, most substrates are soluble in higher concentrations at increased temperatures; however, important exceptions are the gaseous substrates (and products) (see Fig. 2). Of these substrates, oxygen shall be discussed in detail.

On the one hand, the low solubility of oxygen favors anaerobic cultures and no special precautions to eliminate traces of air should be necessary, at least from larger scale bioreactors. On the other hand, both low oxygen solubility and high specific oxygen uptake rates of thermophilic aerobes greatly restrict their application to high performance reactors and (very) dilute suspensions. In Fig. 6, the dependence of necessary k_La on temperature is shown based on the following assumption: A culture with a specific oxygen uptake rate of 30 mMol g^{-1} h^{-1} at a density of 1 g l^{-1} should be maintained at a pO_2 of 50% or 1% of air saturation at the respective temperature; oxygen solubility is assumed to be equal to distilled water under steady state conditions (oxygen uptake rate equals oxygen transfer rate). It is left to the reader to estimate the necessary oxygen transfer capacity of an industrial reactor, when in highly concentrated technical media (instead of water), dense populations

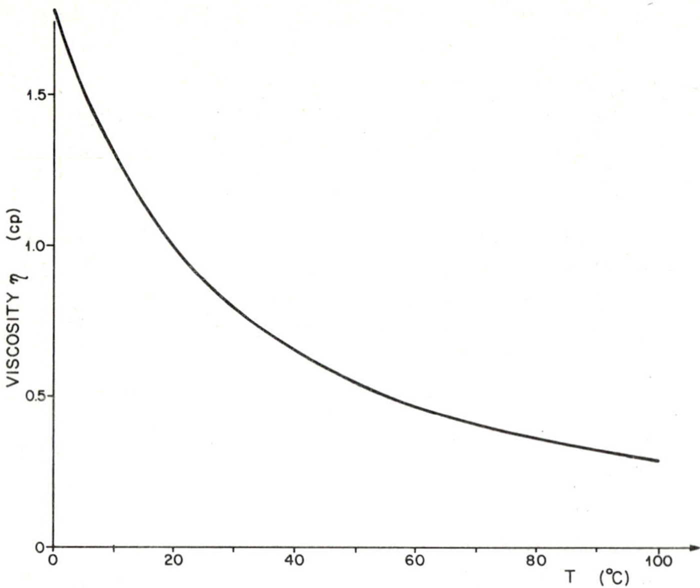

Fig. 5. Temperature dependence of viscosity of pure water. Plotted from [381]

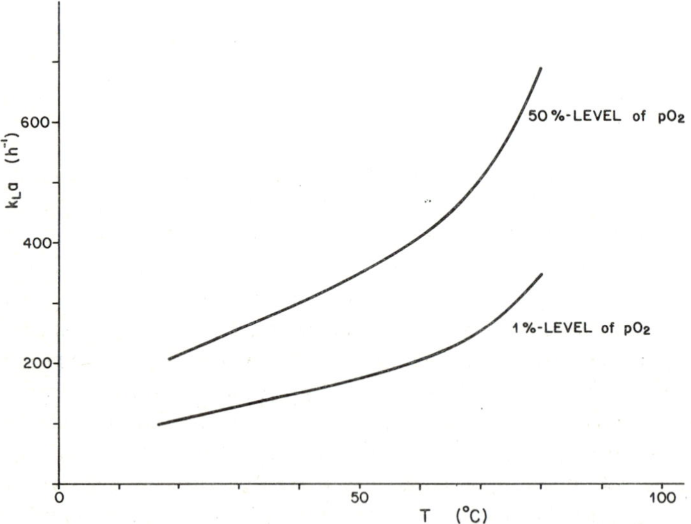

Fig. 6. Temperature dependence of k_La values necessary to maintain a 1% or 50% saturation level of oxygen partial pressure (100% is equivalent to air saturation) under the assumption that an organism with a specific oxygen uptake rate of 30 mMol g^{-1} h^{-1} is cultivated at a cell density of 1 g l^{-1} under steady state conditions where the equivalence:
oxygen transfer rate (OTR) = volumetric oxygen uptake rate (OUR) is valid

(instead of 1 g l^{-1}) must be supplied with sufficient oxygen to take full advantage of the kinetic capacities of thermophiles.

However, the volumetric oxygen transfer rate k_La should also increase with rising temperatures under otherwise constant conditions since this value is affected by changes of both the mass transfer coefficient in the liquid film k_L (viscosity and diffusion are temperature dependent) and of the interfacial area a (surface tension is temperature dependent). In fact, for k_La, at 30 °C, an approximately 15% higher value is reported than at 20 °C and at 10 °C, the k_La is about 15% less than at 20 °C [346]. k_La values for H_2, O_2, and CO_2 determined using the sodium-sulfite oxidation method and estimated from $[k_La]_{O_2}$ at 30 °C and 50 °C, show a more significant influence of temperature: the 20 °C increase doubles the k_La value [347,348]. The biological relevance of sulfite-method derived k_La values shall not be discussed here.

Whether k_La values can be increased to sufficient values by any means of chemistry or engineering remains to be elucidated and checked in biological test systems. Furthermore, critical oxygen partial pressures for thermophiles are not known at present. We have found indications that continuous cultures of *Bacillus caldotenax* or *B. acidocaldarius* at densities less than 1 g l^{-1} but with q_{O_2} values of about 50 mMol g^{-1} h^{-1} were grown at least partially oxygen limited at pO_2 readings of 5 to 10% of air saturation (unpublished results); cultivation temperatures were between 60 and 70 °C, aeration rate and stirrer (impeller with draft tube) speed was 1 to 1.5 vvm and 3000 rpm, respectively.

6.5 Stability of Organisms

Inherent stability or the proper means to stabilize microorganisms are absolute requirements for their technical application. From the data published on instability of characteristic properties of thermophiles, one might assume that besides other kinetic parameters, the probability of mutation or adaptation is also increased. No conclusive answers are possible since the molecular bases of the observed alterations are not clear at present. However, metabolic regulation should — by definition — be reversible. Exact reversibility of physiological and/or kinetic changes has not been reported. On the other hand, during the same period of time, a higher number of generations of the more rapidly growing thermophiles has been observed than of mesophiles; a higher number of mutants would consequently be found even at the same frequency of mutation. But all these speculations do not provide satisfactory explanations as to why the variations of characteristic properties occur either in continuous culture after a relatively low number of generations has been grown or upon repeated subcultivation on solid or liquid media. This fact supports the theory of enrichment of 'specialists' which are either present in the parent culture or arise from mutation due to a selection pressure, the character of which is not known to the experimentator.

Thermus aquaticus could be grown as a pure yellow culture only in batch systems. Upon starting a chemostat, white cell types emerged, increased in number and could overgrow the yellow cells within a few volume changes (independent of dilution rate).

The resulting pure or mixed cultures were either stable or unstable in chemostat as shown by the x-D-diagrams, with no hysteresis (stable population) or with an 'open hysteresis' (change of population) [73,148]. Although the populations were kinetically quite different, on the basis of their biochemical characteristic properties and the typical cross hatched structure of the outer membrane, it was concluded that all the observed populations were *Thermus aquaticus*, too [73,84]. On the other hand, *T. thermophilus* has been found to be stable in chemostat cultures over a period as long as 3 months [72]. *T. thermophilus* has been described as having quite different growth rates (on very similar media): μ_{max} was found to be about 0.15 h^{-1} [152], 0.8 h^{-1} [194], 2.2 h^{-1} [219] and up to 2.7 h^{-1} [72]. The lysine auxotrophy found for *T. thermophilus* [131], support of growth through vitamins and amino acids [28,219,349], and substrate inhibition only at relatively high concentrations [147] could not be verified in chemostat studies [72].

The loss of exoenzyme productivity of *Bacillus subtilis* strain NRRL-B3411 observed in the early phase of a chemostat on different media was interpreted as due to the takeover by mutants [285]. The originally pure (cloned) culture of a thermophilic *Bacillus* producing α-amylase was found to segregate into two types of cells in an early phase of chemostat culture. Both cell types could grow on nutrient broth, but only one, the transparent type, could grow on starch alone. The second type was able to grow in the vicinity of the transparent type on starch agar plates because it could utilize the reducing sugars produced by the readily diffusing α-amylase of the transparent type from starch [73]. The ability of *B. stearothermophilus* strain 1503-4 to produce high amounts of α-amylase has been irreversibly lost, only less than 1% of the originally reported quantity [284] could be detected even upon reisolation of the original rough strain [350].

Clostridium thermocellum, originally requiring more than 0.1% of yeast extract in the medium for growth, could be 'adapted' by a few serial transfers to grow equally well in the presence of only 0.02% yeast extract. A spontaneous mutation of the glucose transport system was assumed to be the reason [11]. However, the glucose transport system of *C. thermocellum* is under metabolic control [351].

For 'thermoadaptation' in the case of *Bacillus caldotenax*, as postulated earlier [104], no indication could be found using continuous culture technique [67].

The problem of instability of thermophilic strains is, of course, a biological one which remains to be clarified, but it obviously is not restricted to thermophiles. It therefore should be regarded instead as a technical problem, and its solution depends on labor efforts and time; mesophilic production strains have not been developed within only a few years; they now have histories of up to several decades of continuous improvement, stabilization and preservation.

6.6 Cultivation Systems

Many microbiologists have tried to carry out their experiments on solidified media or in standing or shaking flasks because these methods are not complicated and are thus easy to use. However, the disadvantages of these systems must be carefully considered. Low heat and mass transfer and incomplete mixing may cause a significant amount of artifacts if the reaction rates of the organisms are high, e.g., even dilute cultures of obligate aerobes may be oxygen limited and/or affected by carbon dioxide. Evaporation must be compensated for either by the addition of sterile water or by calculations. Since pO_2, pH and redox potential cannot be controlled, unknown shifts of these parameters can easily produce incomplete or misinterpreted results, such as 'growth did not cease due to substrate depletion'. On the other hand, the metabolic properties of thermophiles, such as the utilization of special substrates or the excretion of enzymes, are sometimes not expressed in submerged culture [131,278,279] where influences from the relatively inert materials of vessels and supports are not expected. Deficiencies in measurement and control are easily eliminated by using controlled bioreactors. This first step of process development should be made as early as possible, if necessary, in combination with uncontrolled culture systems to find out the significant differences (parameter identification).

Two inherent disadvantages of discontinuous systems are the continuous changes of the chemical conditions and physiological state, and the limited range of observable generations may become significant. Many contaminating organisms may be cultivated and remain undetected since they are not sufficiently diluted out or selected during the short period of growth. Or if adaptation to different concentrations is necessary, as in the case of substrate inhibition, unpredictable, long lag times can occur or the actual growth and turnover rates of substrates and products will never meet the maximal capacity of the organism. The use of continuous culture systems can prevent these problems. Steady states are independent of time and the specific growth rate of the organism can be chosen and preset by the experimentator. These techniques provide a powerful means for measuring the kinetics of fast reactions such as the growth and product formation of thermophiles, for investigating metabolic control, testing the stability of an organism, selecting and detecting improved strains

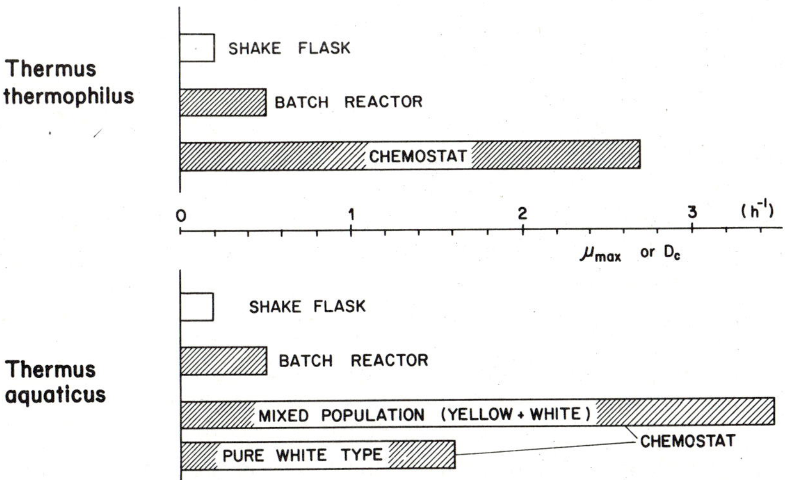

Fig. 7. Dependence of specific growth rate on the culture system. Data from [72,73,148] for *Thermus aquaticus* and *T. thermophilus*. Growth media were: 3 g l^{-1} peptone and 5 g l^{-1} meat extract. Growth conditions: pH 7, 75 °C, in controlled bioreactors: $pO_2 \geqq 80\%$ of air saturation at 75 °C. Only the white type cells of *T. aquaticus* were grown on a defined glucose medium, see [84]

(mutants or new organisms), and last but not least, for developing and optimizing of appropriate media.

Continuous cultivation is advantageous not only for basic research and process development, but also for production purposes if one takes into account that many potentially useful thermophiles are inhibited by either substrates or products. The respective concentrations can be almost arbitrarily chosen, preset and controlled (within the given limits) in a continuous culture. How much this explains why *Thermus* strains can be cultivated in chemostat at more than 5 times greater specific growth rates than in batch systems is not clear (Fig. 7). In any case, these findings undoubtedly indicate that *Thermus* strains show great potential for use in continuous cultivation systems only.

7 Future Prospects

During the last years, in particular, the great technical potential of thermophilic microorganisms has been shown by studies on the formation of thermostable enzymes, immobilization of thermophilic whole cells and enzymes, the production of chemicals, solvents and fuels, and the utilization of abundant renewable substrates by thermophiles. Undoubtedly, the spectrum of products formed and substrates utilized will be increased in the future as new organisms are isolated and/or constructed. From the point of microbial diversity, rapid development and prosperity of thermophilic biotechnology can be expected.

Although many of the expected benefits of thermophilic processes can at present be regarded as realistic, some of them have been overestimated and some disadvantages have probably been underestimated. It would be of practical use to discuss these features from three different points of view: (7.1) technical realization and equipment, (7.2) regulation and metabolic control, and (7.3) genetics.

7.1 Technical Realization of Thermophilic Biotechnology

The most important technical features that distinguish mesophilic processes from thermophilic processes are that the latter have altered conditions for heat and mass transfer, require materials with increased inertness against chemical and physical modifications and aging. The latter problems are likely to be localized and overcome by consequent checks in proper biological test systems and by continuous improvement of cultivation equipment. This is desirable and necessary in general, and fortunately, specific thermophilic problems may enhance progress in this area.

Heat transfer in thermophilic processes is facilitated by a sufficiently high temperature gradient between culture and cooling water, as compared with that required in mesophilic (or even psychrophilic) processes, and must be regarded as a great advantage. Heat is lost because of evaporation at high temperatures, but that loss may be negligible when compared with the advantage of facilitated recovery of volatile products directly from the culture. Moreover, the concentration of such products, which usually have inhibitory effects at relatively low concentrations ('concentration limitation', [14]), can be kept low in the culture. In the case of nonvolatile products, evaporation may be necessarily compensated for by proper means for the same reason of concentration limitation, otherwise evaporation represents a step of product enrichment.

Mass transfer increases with increasing temperature due to decreasing viscosity (and surface tension). Furthermore, many substances are more soluble at higher temperatures than at lower ones. Important exceptions are the gases of biological relevance, O_2, CO_2, H_2, CH_4 (and N_2). Since the temperature-dependent increase of k_La values is by far exceeded by both the higher gas exchange rates and the lower gas solubility, as compared with mesophilic organisms and conditions, aerobic thermophilic processes can only be technically and economically of value either if bioreactors with extremely high transfer performance are available or if the driving gradient is increased by the application of pure oxygen and/or high pressure.

However, extended studies on the oxygen requirements (e.g., critical partial pressure, inhibition by high partial pressure) of aerobic thermophiles and studies on the realization of sufficient oxygen supply (e.g., reactor design, pressurized cultures) are necessary. The presently known nonsporulating thermophilic aerobes are not suitable for a biological test system because they cannot be cultivated in dense cultures (due to substrate inhibition). If it should prove impossible or uneconomical to cultivate very dense cultures of thermophilic aerobes, a solution for the production of thermostable products would be the genetic transfer of the necessary thermophilic genes into proper mesophiles (see Sect. 7.3). Reduction of culture temperature from the optimum down to the minimum is no solution because all the other advantages of high temperature would be diminished and growth rates would be decreased to low levels comparable to or even less than those of mesophiles.

7.2 Regulation and Metabolic Control

Under the general terms regulation and metabolic control, several features must be considered. Quantitative investigations with *Thermus* strains have shown that these organisms grow with both low yield and incomplete substrate utilization. Moreover, most of them are inhibited by relatively low substrate concentrations. Although a great number of patents concerning the production of several thermostable enzymes using *Thermus thermophilus*, *T. flavus*, *T. caldophilus*, etc. have been published (mostly Japanese patents), keeping currently available knowledge in mind, one would expect that these enzymes could be produced more economically and easily using other thermophiles such as *Bacilli*. So far, these organisms have been found to utilize substrates completely, to grow with expected yields (i.e., comparable with mesophilic yields which are reduced only due to maintenance requirements of the organisms), and to produce a wide spectrum of interesting enzymes. However, the high growth temperatures reported for *Thermus* strains have been reported so far only for a few *Bacillus* strains. Further research on thermophilic *Bacilli* can therefore be regarded as promising and should be intensified. An indication of this may be seen in the successful isolation during the past few years of several anaerobic thermophiles with a high potential for rapid growth and rapid product formation.

In general, both anaerobic and aerobic thermophiles cannot be expected to compete for but rather to equally contribute to the future value of thermophilic biotechnology because of the higher metabolic rates of the aerobes and the different product spectrum of the anaerobes [49]. However, thermophilic aerobiosis is more difficult to realize and to scale up than mesophilic aerobiosis due to oxygen transfer problems. Thermophilic anaerobiosis may be easier in practice than mesophilic anaerobiosis due to reduced oxygen solubility.

Concerning 'rate limitation' and 'yield limitation' [14], all the thermophiles investigated have been superior or at least comparable to mesophiles. Moreover, rates and yields may be further improved by optimization of media and culture conditions as well as by the choice of a proper cultivation system, as in continuous culture, where *Thermus* grows significantly more rapidly than in batch (Fig. 7). At present, only 'concentration limitation' seems to be a problem because most of the thermophilic wild types have been shown to be inhibited at lower product or substrate concentrations than comparable mesophilic (production) strains. However, one must bear in mind that the mesophilic production strains have been developed from wild type strains over long periods of time. The same development of tolerant strains among thermophiles can be expected to be achieved in shorter periods of time in the near future, as seems possible due to the successes with *Clostridium thermoaceticum* [189,190,352] and *C. thermocellum* [11,44,45,172].

One may hopefully expect the isolation or construction of a 'superbacterium', which undoubtedly will be a thermophile [11,12]; however, the presently known thermophiles are unequivocally not inferior to mesophilic microorganisms with respect to their physiological capacities, as there are the spectra of substrates utilized or products formed. An expansion of these spectra has been successfully initiated by defined mixed cultures and will be continued in the future. The unrestricted physiology of most thermophiles will possibly be a basis for the solution of specific waste treatment problems that cannot be economically solved using mesophiles because of their low metabolic rates.

7.3 Genetic Means

Genetics will play an important role in the future biotechnology of thermophiles. First, it must be elucidated whether thermophiles are really subject to higher mutation frequencies, as has been indicated by several observed changes of characteristic qualitative or quantitative properties; if so, the causes of mutation and how a biotechnologist can cope with the resulting problems must be determined. Serious investigations are necessary in the near future to answer these questions.

Second, the genetic constructions of thermophilic genes in mesophiles should either facilitate product recovery or allow production on a technical scale. The transformation of genes from thermophiles into mesophiles and the expression of a thermophilic gene product has been shown in several examples: α-amylase from the thermophile V2 transformed into *Bacillus subtilis* is expressed with nearly the fully intact thermostability of the enzyme of the parent strain [272,353]. 3-isopropylmalate dehydrogenase from *Thermus thermophilus* HB8 has been cloned in *Escherichia coli* using pBR322 as a vector. Moreover, *E. coli* cells carrying the recombinant plasmid (pHB2) produced 7 times more of the thermostable enzyme than the parent *Thermus thermophilus* and a 4 fold purification of the enzyme could be accomplished by simple heat treatment of the crude extract of the plasmid carrying *E. coli* cells [354]. The hybrid plasmid pBR322-T leu carrying *T. thermophilus* HB27 DNA including structural gene and promotor coded for a 3-isopropylmalate dehydrogenase with an optimal temperature of 80 °C in the host *E. coli*. However, the thermostability of the gene product seems to be restricted and gradually lost with time [355].

These few successful experiments clearly show that recovery of thermostable products formed by mesophiles (or psychrophiles) is easy through simple enrichment by heat treatment. If products cannot be economically produced using the thermophile for reasons such as insufficient oxygen transfer or very low yields, the expression of the desired gene product by a (easily culturable) mesophile is at least a likely alternative, if not a general solution, to such problems. In order to obtain a maximum of safety, one should consider the use of plasmids isolated from thermophiles [356].

After summarizing the present knowledge of thermophiles, one must admit that in spite of all the still-unsolved technical, physiological, and genetic problems, the potential usefulness of (extreme) thermophilic microorganisms is impressive. Further efforts have to be undertaken to screen for 'new organisms' and for new products that can be more reasonably produced with thermophiles. Physiology and metabolic control of thermophiles must be studied in further detail and new fields of application for biocatalysts that are currently interesting only from an academic point of view have to be investigated. The stability of strains and their tolerance towards higher substrate and product concentrations must be increased. Process design, including high performance reactors and product recovery, as well as improved measurement and control of bioprocesses should be stimulated again to make further progress. Proper biological test systems would be necessary to prove the correct scale-up criteria, and continuous cultivation systems would be of increased advantage over discontinuous systems. Thermophily is not only superior to mesophily in the production of many products and in the removal of several substrates, but should undoubtedly give new stimuli for the development of biotechnology in general.

8 Note Added in Proof

Important results, published since initial completion of this manuscript, should be added because they affirm statements made in the original text or bring new aspects into consideration.

Sulfur-hydrogen autotrophic disc-shaped prokaryotic organisms have been isolated from a submarine solfatara that grew optimally at 105 °C with a doubling time of 110 min ($\mu = 0.38\ h^{-1}$) [383]. The minimal growth temperature of these strict anaerobes was about 80 °C in synthetic media. This further observation of growth at temperatures greater than 100 °C clearly supports the hypothesis that microbial life is not limited by temperature, but by the liquid state of the environment (medium).

The analytical or commercial value of the unusual alcohol dehydrogenases produced by either *Thermoanaerobium brockii* or *Clostridium thermohydrosulfuricum* (see Sect. 5.2.3) has been emphasized by a new patent [384]. The optical purity of the reduction product could be upgraded from about 70 to about 80% (in the case of 2-pentanone reduction) by lowering the temperature to 25 °C and by diluting the educt down to 4 g l^{-1}. The net reduction of ketones could be increased by addition of hydrogen if growing cells were used as catalyst.

Successful expression of thermophilic genes in *Escherichia coli* has been reported for several enzymatic activities encoded by *Clostridium thermocellum* genes [385]. However, the expression of endoglucanase in the host was only about 30% that in the parent strain (enzyme test at 60 °C). The normally extracellular enzyme of *C. thermocellum* is intracellular in *Escherichia coli*. Unfortunately, no data are given on the stabilities of the cloned enzymes.

A kinetic study with *Thermothrix thiopara* using continuous culture also revealed a higher μ_{max} value (0.58 h^{-1}) [386] than earlier studies, which used batch techniques (see Tab. 2). Interestingly, the specific maintenance rate of this organism was found to decrease with an increasing growth temperature: a = 0.41, 0.11, and 0.15 h^{-1} at 65, 70, and 75 °C, respectively. The true yield (Y_G) decreased in the same way: 41, 24, and 19 g mol^{-1} at 65, 70, and 75 °C (compare Sect. 4.5). However, in spite of the high maintenance requirements, the growth yield was higher than that obtained with any known mesophilic sulfur-chemoautotroph.

Methane production from volatile fatty acids using mixed cultures originating from mesophilic sewage sludge has been shown to occur at higher rates at temperatures between 55 and up to 85 °C than at mesophilic temperatures [387]. However, CO_2 caused the same degree of inhibition and the methane yield was even lower.

An interesting application of *Sulfolobus acidocaldarius* in the removal of sulfur compounds from coal at 75 °C and pH of around 2.5 has been reported [388, 389]. The efficiency of the process greatly depended on the concentrations of yeast extract and Fe in the media, as well as on the sulfur content of the coal and its particle size distribution (specific surface area). A 50% removal of initial total sulfur present could be achieved.

9 Acknowledgements

I am grateful to Prof. A. Fiechter for his encouragement and appreciate his continuous interest in thermophilic biotechnology research. I thank Lee Anne Clark for the preparation of the manuscript.

10 Symbols

a	h^{-1}	Specific maintenance rate according to model of Marr et al. [205]
D	h^{-1}	Dilution rate
D_c	h^{-1}	Critical dilution rate
E_a	J mol^{-1}	Activation energy in Arrhenius equation
k		Rate constant (general)

k_∞	same as k	Arrhenius constant
k_d	h^{-1}	Specific death rate
k_La	h^{-1}	Volumetric gas transfer rate
K_s	$g\,l^{-1}$	Saturation constant according to growth model of Monod
K_M	$g\,l^{-1}$	Saturation constant according to enzyme kinetics after Michaelis-Menten
m_S	h^{-1}	Maintenance coefficient for substrate according to model of Pirt [206,207]
M_{O_2}	$mMol\,g^{-1}\,h^{-1}$	Maintenance coefficient for O_2 (oxygen) according to model of Pirt
pH_2	Pa	Partial pressure of hydrogen
pO_2	Pa	Partial pressure of oxygen
q	h^{-1}	Specific metabolic rate:
q_{O_2}	$mMol\,g^{-1}\,h^{-1}$	Specific oxygen uptake rate
q_S	h^{-1}	Specific substrate uptake rate
q_P	h^{-1}	Specific product formation rate
r_S	$g\,l^{-1}\,h^{-1}$	Rate of substrate consumption (absolute)
r_X	$g\,l^{-1}\,h^{-1}$	Rate of biomass production (absolute)
s	$g\,l^{-1}$	Substrate concentration
s_0	$g\,l^{-1}$	Substrate concentration in sterile medium
t_d	h (min)	Doubling time
x	$g\,l^{-1}$	Biomass concentration
Y	$g\,g^{-1}$	Yield coefficient. If not specified by indices $Y_{X/S}$ is assumed:
$Y_{X/S}$	$g\,g^{-1}$	Yield: biomass produced per substrate consumed
$Y_{P/S}$	$g\,g^{-1}$	Yield: product formed per substrate consumed
$Y_{X/P}$	$g\,g^{-1}$	Yield: biomass produced per product formed
$Y_{P/X}$	$g\,g^{-1}$	Yield: product formed per biomass produced
Y_G	$g\,g^{-1}$	Yield corrected for maintenance according to model of Pirt
η	Pa s (poise)	Viscosity (absolute)
μ	h^{-1}	Specific growth rate
μ_{max}	h^{-1}	Maximal specific growth rate

11 References

1. Bott, T. L., Brock, T. D.: Nature *164*, 1411 (1969)
2. Brock, T. D.: Science *158*, 1012 (1967)
3. Brock, T. D., Brock, M. L.: Nature *233*, 494 (1971)
4. Heinen, W., Lauwers, A. M.: Arch. Microbiol. *129*, 127 (1981)
5. Baross, J. A., Lilley, M. D., Gordon, L. I.: Nature *298(5872)*, 366 (1982)
6. Amelunxen, R. E., Murdock, A. L.: Crit. Rev. Microbiol. *6*, 343 (1978)
7. Ljungdahl, L. G.: Adv. Microb. Physiol. *19*, 149 (1979)
8. Daniel, R. M., Cowan, D. A., Morgan, H. W.: Chem. New Zealand *45(3)*, 94 (1981)
9. Ljungdahl, L. G., Carreira, L., Wiegel, J.: The Ekman-days 4, IV23 (1981)
10. deRosa, M., Gambacorta, A., Nicolaus, B., Buonocore, V., Poerio, E.: Biotechnol. Letters *2(1)*, 29 (1980)
11. Wiegel, J.: Experientia *36*, 1434 (1980)
12. Zeikus, J. G.: Enzyme Microb. Technol. *1*, 243 (1979)
13. Zeikus, J. G., Ben-Bassat, A., Hegge, P. W.: J. Bacteriol. *143(1)*, 432 (1980)
14. Zeikus, J. G., Ben-Bassat, A., Ng, T. K., Lamed, R. J.: Basic Life Sci. *18*, 441 (1981)

15. Woese, C. R.: Spektrum d. Wissenschaft *8*, 75 (1981)
16. Woese, C. R., Magrum, L. J., Fox, G. E.: J. Mol. Evol. *11*, 245 (1978)
17. Zillig, W., Tu, J., Holz, I.: Nature *293*, 85 (1981)
18. Brock, T. D.: Thermophilic microorganisms and life at high temperatures. (Starr, M. P. ed.), Berlin, Heidelberg, New York: Springer 1978
19. Langworthy, T. A.: Life Sci. Res. Rep. *13*, 489 (1979)
20. Allen, M. B.: Bacteriol. Rev. *17*, 125 (1953)
21. Aragno, M.: Responses of microorganisms to temperature, in: Encyclopedia of plant physiology, new series, Vol 12A; Physiological plant ecology I, p. 339, (Lange, O. L., Nobel, P. S., Osmond, C. B., Ziegler, H., eds.) Berlin, Heidelberg, New York: Springer 1981
22. Ben-Bassat, A., Zeikus, J. G.: Arch. Microbiol. *128*, 365 (1981)
23. Brannan, D. K., Caldwell, D. E.: Appl. Environm. Microbiol. *40(2)*, 211 (1980)
24. Brock, T. D., Freeze, H.: J. Bacteriol. *98(1)*, 289 (1969)
25. Fliermans, C. B., Brock, T. D.: ibid. *111(2)*, 343 (1972)
26. Jackson, T. J., Ramaley, R. F., Meinschein, W. G.: Int. J. Syst. Bacteriol. *23(1)*, 28 (1973)
27. Loginova, L. G., Userbaeva, G. B., Egorova, L. A.: Mikrobiologiya *47(6)*, 1081 (1978)
28. Oshima, T., Imahori, K.: J. Gen. Appl. Microbiol. *17*, 513 (1971)
29. Pask-Hughes, R. A., Williams, R. A. D.: J. Gen. Microbiol. *102*, 375 (1977)
30. Ramaley, R. F., Bitzinger, K.: Appl. Microbiol. *30(1)*, 152 (1975)
31. Ramaley, R. F., Bitzinger, K., Carroll, R. M., Wilson, R. B.: Int. J. Syst. Bacteriol. *25(4)*, 357 (1975)
32. Ramaley, R. F., Hixon, J.: J. Bacteriol. *103(2)*, 527 (1970)
33. Rozanova, E. P., Khudyakova, A. I.: Mikrobiologiya 43*(6)*, 1069 (1974)
34. Saiki, T., Kimura, R., Arima, K.: Agric. Biol. Chem. *36(13)*, 2357 (1972)
35. Wiegel, J., Ljungdahl, L. G., Rawson, J. R.: J. Bacteriol. *139(3)*, 800 (1979)
36. Zeikus, J. G., Hegge, P. W., Anderson, M. A.: Arch. Microbiol. *122*, 41 (1979)
37. Zeikus, J. G., Wolfe, R. S.: J. Bacteriol. *109(2)*, 707 (1972)
38. Alexander, J. K., Connors, R., Yamamoto, N.: Production of liquid fuels from cellulose by combined saccharification-fermentation or cocultivation of *Clostridia*, in: Advances in Biotechnology, II; Fuels, chemicals, foods and waste treatment, p. 125, (Moo-Young, M., ed.) Oxford: Pergamon Press 1981
39. Gordon, J., Cooney, C. L.: Fermentation production of sugars from cellulose by *Clostridium thermocellum*, in: Advances in Biotechnology, II; Fuels, chemicals, foods and waste treatment, p. 15, (Moo-Young, M., ed.) Oxford: Pergamon Press 1981
40. Ljungdahl, L. G., Bryant, F., Carreira, L., Saiki, T., Wiegel, J.: Basic Life Sci. *18*, 397 (1981)
41. Ng, T. K., Weimer, P. J., Zeikus, J. G.: Arch. Microbiol. *114*, 1 (1977)
42. Ng, T. K., Zeikus, J. G.: Appl. Environm. Microbiol. *42(2)*, 231 (1981)
43. Weimer, P. J., Zeikus, J. G.: ibid. *33(2)*, 289 (1977)
44. Avgerinos, G. C., Fang, H. Y., Biocic, I., Wang, D. I. C.: A novel, single-step microbial conversion of cellulosic biomass to ethanol, in: Advances in Biotechnology, II; Fuels, chemicals, foods and waste treatment, p. 119, (Moo-Young, M., ed.) Oxford: Pergamon Press 1981
45. Herrero, A., Gomez, R. F.: Appl. Environm. Microbiol. *40(3)*, 571 (1980)
46. Lamed, R., Zeikus, J. G.: J. Bacteriol. *144(2)*, 569 (1980)
47. Lamed, R. J., Keinan, E., Zeikus, J. G.: Enzyme Microb. Technol. *3*, 144 (1981)
48. Saddler, J. N., Chan, M. K. H., Luis-Seize, G.: Biotechnol. Letters *3(6)*, 321 (1981)
49. Zeikus, J. G.: Ann. Rev. Microbiol. *34*, 423 (1980)
50. Zertuche, L., Zall, R. R.: Biotech. Bioeng. *24*, 57 (1982)
51. Bochem, H. P., Schoberth, S. M., Sprey, B., Wengler, P.: Can. J. Microbiol. *28*, 500 (1982)
52. Cooney, C. L., Wise, D. L.: Biotech. Bioeng. *17*, 1119 (1975)
53. Hashimoto, A. G., Chen, Y. R., Prior, R. L.: J. Soil. Water Conserv. *34*, 16 (1979)
54. Schoenheit, P., Moll, J., Thauer, R. K.: Arch. Microbiol. *127*, 59 (1980)
55. Stetter, K. O., Thomm, M., Winter, J., Wildgruber, G., Huber, H., Zillig, W., Janecovic, D., Koenig, H., Palm, P., Wunderl, S.: Zbl. Bakt. Hyg. I. Abt. Orig. Ċ *2*, 166 (1981)
56. Taylor, G. T.: Prog. Ind. Microbiol. *16*, 231 (1982)
57. Taylor, G. T., Pirt, S. J.: Arch. Microbiol. *113*, 17 (1977)
58. Varel, V. H., Hashimoto, A. G., Chen, Y. R.: Appl. Environm. Microbiol. *40(2)*, 217 (1980)
59. Varel, V. H., Isaacson, H. R., Bryant, M. P.: ibid. *33(2)*, 298 (1977)
60. Zeikus, J. G.: Bacteriol. Rev. *41*, 514 (1977)

61. Brierley, C. L.: Sci. Americ. *247(2)*, 42 (1982)
62. Brierley, J. A., Lockwood, S. J.: FEMS Microbiol. Letters *2*, 163 (1977)
63. Brierley, J. A., Norris, P. R., Kelly, D. P., Le Roux, N. W.: Eur. J. Appl. Microbiol. Biotechnol. *5*, 291 (1978)
64. Norris, P. R., Brierley, J. A., Kelly, D. P.: FEMS Microbiol. Letters 7, 119 (1980)
65. Heinen, U. J., Heinen, W.: Arch. Microbiol. *82*, 1 (1972)
66. Heinen, W.: ibid *76*, 2 (1971)
67. Kuhn, H. J., Cometta, S., Fiechter, A.: Eur. J. Appl. Microbiol. Biotechnol. *10*, 303 (1980)
68. Bergey's manual of determinative bacteriology, (Buchanan, R. E., Gibbons, N. E., eds.), Baltimore: Williams Wilkins Comp. 1975[8]
69. Wolf, J., Sharp, R. J.: Taxonomic and related aspects of thermophiles within the genus *Bacillus*, in: The aerobic endospore-forming bacteria: Classification and identification, p. 251, (Berkeley, R. C. W., Goodfellow, M., eds.), New York: Acad. Press (1981)
70. Cometta, S., Sonnleitner, B., Sidler, W., Fiechter, A.: Eur. J. Appl. Microbiol. Biotechnol. *16*, 151 (1982)
71. Lungershausen, R.: Untersuchungen zur Taxonomie und Wachstumsphysiologie extrem thermophiler Bakterien. PhD thesis, Goettingen (1978)
72. Sonnleitner, B., Cometta, S., Fiechter, A.: Eur. J. Appl. Microbiol. Biotechnol. *15*, 75 (1982)
73. Sonnleitner, B., Cometta, S., Fiechter, A.: 4th Int. Sympos. Genetics Industrial Microorganisms (1982)
74. Loginova, L. G., Egorova, L. A.: New forms of thermophilic bacteria. Reviewed by Kvasnikov, E. Mikrobiologiya *47(4)*, 773 (1978)
75. Gottschalk, G., Andreesen, J. R., Hippe, H.: The genus *Clostridium* (nonmedical aspects), in: The prokaryotes, p. 1767, (Starr, M. P., Stolp, H., Trueper, H. G., Balows, A., Schlegel, H. G., eds.), Berlin, Heidelberg, New York: Springer 1981
76. Klaushofer, H., Hollaus, F.: Z. Zuckerind. *20(9)*, 465 (1970)
77. Sokal, R., Sneath, P. H. A.: Principles of numerical taxonomy; London: Freeman Comp. 1963
78. Wolf, J., Barker, A. N.: The genus *Bacillus*: Aids to the identification of its species, in: Identification methods for microbiologists, part B, p. 93, (Gibbs, B. M., Shapton, D. A., eds.), New York: Acad. Press 1968
79. Walker, P. D., Wolf, J.: The taxonomy of *Bacillus stearothermophilus*, in: Spore Research, p. 247, (Baker, A. N., ed.), New York: Acad. Press 1971
80. Prangishvilli, D., Zillig, W., Gierl, A., Biesert, L., Holz, I.: Eur. J. Biochem. *122*, 471 (1982)
81. Zillig, W., Stetter, K. O., Wunderl, S., Schulz, W., Priess, H., Scholz, I.: Arch. Microbiol. *125*, 259 (1980)
82. Moser, A.: Bioprozesstechnik; Berlin, Heidelberg, New York: Springer 1981
83. Castenholz, R. W.: Life Sci. Res. Rep. *13*, 373 (1979)
84. Cometta, S., Sonnleitner, B., Fiechter, A.: Eur. J. Appl. Microbiol. Biotechnol. *15*, 69 (1982)
85. Amelunxen, R. E., Murdock, A. L.: Develop. Ind. Microbiol. *18*, 233 (1976)
86. Amelunxen, R. E., Murdock, A. L.: Microbial life at high temperatures: Mechanisms and molecular aspects, in: Microbial life in extreme environments, p. 217, (Kushner, D. J., ed.), New York: Acad. Press 1978
87. Argos, P., Rossmann, M. G., Grau, U. M., Zuber, H., Frank, G., Tratschin, J. D.: Biochemistry *18*, 5698 (1979)
88. Degryse, E.: Experientia Suppl. *26*, 401 (1976)
89. Ljungdahl, L. G., Sherod, D.: ibid. *26*, 147 (1976)
90. Neurath, H.: ibid. *26*, 412 (1976)
91. Oshima, T.: Life Sci. Res. Rep. *13*, 455 (1979)
92. Oshima, T., Sakaki, Y., Wakayama, N., Watanabe, K., Ohashi, Z., Nishimura, S.: Experientia Suppl. *26*, 317 (1976)
93. Singleton, R.: 76th Ann. Meet. Americ. Soc. Microbiol. (1976)
94. Singleton, R., Amelunxen, R. E.: Bacteriol. Rev. *37*, 320 (1973)

95. Tansey, M. R., Brock, T. D.: Microbial life at high temperatures: Ecological aspects, in: Microbial life in extreme environments, p. 159, (Kushner, D. J., ed.), New York: Acad. Press 1978
96. Walker, J. E.: Proc. FEBS Meet. *52*, 211 (1978)
97. Williams, R. A. D.: Sci. Prog. Oxf. *62*, 373 (1975)
98. Zuber, H., Aminopeptidases from thermophilic microorganisms in: Peptides 1969, p. 310, North-Holland Publ. Comp. 1971
99. Zuber, H.: Comparative studies of thermophilic and mesophilic enzymes: Objectives, problems, result., in: Biochemistry of thermophily, p. 267, (Friedman, S. M., ed.), New York: Acad. Press 1978
100. Zuber, H.: Life Sci. Res. Rep. *13*, 393 (1979)
101. Zuber, H.: Chemie in unserer Zeit *6*, 165 (1979)
102. Zuber, H.: Structure and function of thermophilic enzymes, in: 32. Coll. der Ges. fuer Biologische Chemie; Structural and functional aspects of enzyme catalysis, (Eggerer, H., Huber, R., eds.), Berlin, Heidelberg, New York: Springer 1981
103. Zuber, H.: Basic Life Sciences *18*, 499 (1981)
104. Haberstich, H. U., Zuber, H.: Arch. Microbiol. *98*, 275 (1974)
105. Frank, G., Haberstich, H. U., Schaer, H. P., Tratschin, J. D., Zuber, H.: Experientia Suppl. *26*, 375 (1976)
106. Bauman, A. J., Simmonds, P. G.: J. Bacteriol. *98(2)*, 528 (1969)
107. Esser, A. F.: Life Sci. Res Rep. *13*, 433 (1979)
108. Langworthy, T. A.: Biochim. Biophys. Acta *487*, 37 (1977)
109. Langworthy, T. A.: J. Bacteriol. *130(3)*, 1326 (1977)
110. Langworthy, T. A.: Life Sci. Res. Rep. *13*, 417 (1979)
111. Langworthy, T. A., Mayberry, W. R.: Biochim. Biophys. Acta *431*, 570 (1976)
112. Langworthy, T. A., Mayberry, W. R., Smith, P. F.: ibid. *431*, 550 (1976)
113. Oshima, M.: Structure and function of membrane lipids in thermophilic bacteria, in: Biochemistry of thermophily, p. 1, (Friedman, S. M., ed.), New York: Acad. Press 1978
114. Oshima, M., Yamakawa, T.: Biochemistry *13(6)*, 1140 (1974)
115. Poralla, K., Kannenberg. E., Blume, A.: FEBS Letters *113(1)*, 107 (1980)
116. Pask-Hughes, R. A., Shaw, N.: J. Bacteriol. *149(1)*, 54 (1982)
117. Pask-Hughes, R. A., Williams, R. A. D.: J. Gen. Microbiol. *107*, 65 (1978)
118. deRosa, M., Gambacorta, A., Nicolaus, B., Sodano, S., Bu'lock, J. D.: Phytochemistry *19*, 833 (1980)
119. deRosa, M., deRosa, S., Gambacorta, A., Bu'lock, J. D.: ibid. *19*, 249 (1980)
120. Searcy, D. G., Whatley, F. R.: Zbl. Bakt. Hyg. I. Abt. Orig. C *3*, 245 (1982)
121. Wakayama, N., Oshima, T.: J. Biochem. *83*, 1687 (1978)
122. Weerkamp, A., Heinen, W.: Arch. Microbiol. *81*, 350 (1972)
123. Friedman, S. M., Weinstein, I. B.: Biochem. Biophys. Res. Comm. *21*, 339 (1965)
124. Johnson, J. R.: Life Sci. Res. Rep. *13*, 471 (1979)
125. Lindsay, J. A., Creaser, E. H.: Nature *255*, 650 (1975)
126. Lindsay, J. A., Creaser, E. H.: Experientia Suppl. *26*, 391 (1976)
127. Mitchell, R.: Quart. Rev. Biol. *49*, 229 (1974)
128. Stenesh, J.: Experientia Suppl. *26*, 85 (1976)
129. Hill, E. O.: The genus *Clostridium* (medical aspects), in: The prokaryotes, p. 1756, (Starr, M. P., Stolp, H., Trueper, H. G., Balows, A., Schlegel, H. G., eds.), Berlin, Heidelberg, New York: Springer 1981
130. Norris, J. R., Berkeley, R. C. W., Logan, N. A., O'Donnell, A. G.: The genera *Bacillus* and *Sporolactobacillus*, in: The prokaryotes, p. 1711, (Starr, M. P., Stolp, H., Trueper, H. G., Balows, A., Schlegel, H. G., eds.), Berlin, Heidelberg, New York: Springer 1981
131. Degryse, E., Glansdorff, N., Pierard, A.: Arch. Microbiol. *117*, 189 (1978)
132. Egorova, L. A., Bobyk, M. A., Loginova, L. G., Kulaev, I. S.: Mikrobiologiya *50(4)*, 603 (1981)
133. Jackson, T. J., Ramaley, R. F., Meinschein, W. G.: Arch. Microbiol. *88*, 127 (1973)
134. Loginova, L. G., Bogdanova, T. G., Seregina, L. M.: Mikrobiologiya *44(3)*, 489 (1975)
135. Morgan, H. W., Daniel, R. M., Cowan, D. A., Hickey, C. W.: Thermo-stable micro-organism and proteolytic enzyme prepared therefrom, and process for the preparation of

the micro-organism and the proteolytic enzyme. Eur. Patent Applic. # 80302743.2, Publ. # 0 024 182 (1981)
136. Pask-Hughes, R., Williams, R. A. D.: J. Gen. Microbiol. *88*, 321 (1975)
137. Taguchi, H., Yamashita, M., Matsuzawa, H., Ohta, T.: J. Biochem. *91*, 1343 (1982)
138. Wiegel, J., Ljungdahl, L. G.: Arch. Microbiol. *128*, 343 (1981)
139. Brock, T. D.: Ecology *48*, 566 (1967)
140. Brock, T. D., Brock, M. L.: J. Appl. Bacteriol. *31*, 54 (1968)
141. Castenholz, R. W.: Bacteriol. Rev. *33*, 476 (1969)
142. Hoell, K.: Berichte aus der Forschungsstelle Nedri As *6*, 1 (1971)
143. White, D. E., Hem, J. D., Waring, G. A.: Chemical composition of surface waters, in: Data of Geochemistry (Fleischer, M. ed.), US Governm. Print Office: Chapter F 1963[6]
144. Doemel, W. N., Brock, T. D.: Appl. Environm. Microbiol. *34(4)*, 433 (1977)
145. Brock, T. D., Brock, M. L.: J. Bacteriol. *95(3)*, 811 (1968)
146. Mosser, J. L., Bohool, B. B., Brock, T. D.: ibid. *118(3)*, 1075 (1974)
147. Brock, T. D.: Extreme thermophiles of the genera *Thermus* and *Sulfolobus*, in: The prokaryotes, p. 978, (Starr, M. P., Stolp, H., Trueper, H. G., Balows, A., Schlegel, H. G., eds.), Berlin, Heidelberg, New York: Springer 1981
148. Cometta, S.: Zur Biologie des caldoaktiven asporogenen *Thermus aquaticus*. PhD thesis, Swiss Federal Inst. of Technology, Nr. 7063 (1982)
149. Baker, H., Frank, O., Pasher, I., Black, B., Hutner, S. H., Sobotka, H.: Can. J. Microbiol. *6*, 557 (1960)
150. Degryse, E., Glansdorff, N.: Arch. Microbiol. *129*, 173 (1981)
151. Kenkel, T., Trela, J. M.: J. Bacteriol. *140(2)*, 543 (1979)
152. McKay, A., Quilter, J., Jones, C. W.: Arch. Microbiol. *131*, 43 (1982)
153. Kuhn, H. J., Friederich, U., Fiechter, A.: Eur. J. Appl. Microbiol. Biotechnol. *6*, 341 (1979)
154. leRoux, N. W., Wakerley, D. S., Hunt, S. D.: J. Gen. Microbiol. *100*, 197 (1977)
155. Johnson, E. A., Madia, A., Demain, A. L.: Appl. Environm. Microbiol. *41(4)*, 1060 (1981)
156. Zillig, W., Stetter, K. O., Schaefer, W., Janekovic, D., Wunderl, S., Holz, I., Palm, P.: Zbl. Bakt. Hyg. I. Abt. Orig. C *2*, 205 (1981)
157. Darland, G., Brock, T. D.: J. Gen. Microbiol. *67*, 9 (1971)
158. Handley, P. S., Knight, D. G.: Arch. Microbiol. *102*, 155 (1975)
159. deRosa, M., Gambacorta, A., Minale, L., Bu'lock, J. D.: Biochem. J. *128*, 751 (1972)
160. Uchino, F., Katano, T.: Agric. Biol. Chem. *45(4)*, 1005 (1981)
161. Oshima, T., Arakawa, H., Baba, M.: J. Biochem. *81*, 1107 (1977)
162. Schenk, A., Aragno, M.: J. Gen. Microbiol. *115*, 333 (1979)
163. Brierley, C. L.: Develop. Ind. Microbiol. *18*, 273 (1977)
164. Brierley, C. L., Murr, L. E.: Science *170*, 488 (1973)
165. McClure, M. A., Wyckoff, R. W. G.: J. Gen. Microbiol. *128*, 433 (1982)
166. Torma, A. E., Bosecker, K.: Prog. Ind. Microbiol. *16*, 77 (1982)
167. Darland, G., Brock, T. D., Samsonoff, W., Conti, S. F.: Science *170*, 1416 (1970)
168. Hollaender, R.: Forum Mikrobiol. *3*, 160 (1980)
169. Langworthy, T. A.: The Mycoplasmas *1*, 495 (1979)
170. Wiegel, J., Ljungdahl, L. G.: Ethanol as fermentation product of extreme thermophilic, anaerobic bacteria, in: Technische Mikrobiologie, 4. Symp. Berlin, p. 117, (Dellweg, H. ed.) 1979
171. Ng, T. K., Ben-Bassat, A., Zeikus, J. G.: Appl. Environm. Microbiol. *41(6)*, 1337 (1981)
172. Herrero, A. A., Gomez, R. F.: Inhibition of *Clostridium thermocellum* by fermentation products and related compounds, in: Advances in Biotechnology, II; Fuels, chemicals, foods and waste treatment, p. 213, (Moo-Young, M., ed.), Oxford: Pergamon Press 1981
173. Kushner, D. J.: Life in high salt and solute concentrations: Halophilic bacteria, in: Microbial life in extreme environments, p. 318, (Kushner, D. J., ed.), New York: Acad. Press 1978
174. Oshima, T., Imahori, K.: Int. J. Syst. Bacteriol. *24(1)*, 102 (1974)
175. Brock, T. D., Cook, S., Petersen, S., Mosser, J. L.: Geochim. Cosmochim. Acta *40*, 493 (1976)
176. Perski, H. J., Moll, J., Thauer, R. K.: Arch. Microbiol. *130*, 319 (1981)
177. Huber, H., Thomm, M., Koenig, H., Thies, G., Stetter, K. O.: ibid. *132*, 47 (1982)
178. Allen, M. B.: ibid. *32*, 270 (1959)
179. Dieckert, G., Konheiser, U., Piechulla, K., Thauer, R. K.: J. Bacteriol. *148(2)*, 459 (1981)
180. Dieckert, G., Thauer, R. K.: FEMS Microbiol. Letters *7*, 187 (1980)

181. Dieckert, G., Weber, B., Thauer, R. K.: Arch. Microbiol. *127*, 273 (1980)
182. Drake, H. L.: J. Bacteriol. *149(2)*, 561 (1982)
183. Schoenheit, P., Moll, J., Thauer, R. K.: Arch. Microbiol. *123*, 105 (1979)
184. Vogels, G. D., Keltjens, J. T., Hutten, T. J., van der Drift, C.: Zbl. Bakt. Hyg. I. Abt. Orig. C *3*, 258 (1982)
185. Ljungdahl, L. G.: Trace elements and the synthesis of acetate by *Clostridium thermoaceticum*, in: Sci. Sci., p. 89, (Kageyama, M., Nakamura, K., Oshima, T., eds.) Jap. Sci. Soc. Press Tokyo 1981
186. Saiki, T., Shackleford, G., Ljungdahl, L. G.: Selenium Biol. Med. (Proc. Int. Symp.), p. 220 (1981)
187. Brock, T. D., Darland, G. K.: Science *169*, 1316 (1970)
188. Brune, G., Schoberth, S. M., Sahm, H.: Proc. Biochem. *17*, 20 (1982)
189. Schwartz, A. D., Keller, F. A.: Acetic acid by fermentation. Europ. Patent Applic. # 81104811.5, Publ. # 0 043 071 (1982)
190. Schwartz, R. D., Keller, F. A.: Appl. Environm. Microbiol. *43(6)*, 1385 (1982)
191. Vogels, G. D.: Antonie v. Leeuwenhoek *45*, 347 (1979)
192. Hungate, R. E.: Methods Microbiol. *3B*, 117 (1969)
193. Morris, J. G.: Adv. Microb. Physiol. *12*, 169 (1975)
194. Lienert, I., Richter, D.: J. Gen. Microbiol. *123*, 383 (1981)
195. Bohool, B. B., Brock, T. D.: Arch. Microbiol. *97*, 181 (1974)
196. Atkinson, A., Ellwood, D. C., Evans, C. G. T., Yeo, R. G.: Biotech. Bioeng. *17*, 1375 (1975)
197. Klaushofer, H., Parkkinen, E.: Z. Zuckerind. *15(8)*, 445 (1965)
198. Bryant, F., Ljungdahl, L. G.: Biochem. Biophys. Res. Comm. *100(2)*, 793 (1981)
199. Schwartz, R. D., Keller, F. A.: Appl. Environm. Microbiol. *43(1)*, 117 (1982)
200. Brandis, A., Thauer, R. K., Stetter, K. O.: Zbl. Bakt. Hyg. I. Abt. Orig. C *2*, 311 (1981)
201. Kenealy, W. R., Thompson, T. E., Schubert, K. R., Zeikus, J. G.: J. Bacteriol. *150(3)*, 1357 (1982)
202. Norris, P. R., Kelly, D. P.: FEMS Microbiol. Letters *4*, 143 (1978)
203. Chakraborti, N., Murr, L. E.: Biotech. Bioeng. *21*, 1685 (1979)
204. Golovacheva, R. S., Karavaiko, G. I.: Mikrobiologiya *47(5)*, 815 (1978)
205. Marr, A. G., Nilson, E. H., Clark, D. J.: Ann. N. Y. Acad. Sci. *102*, 536 (1963)
206. Pirt, S. J.: Proc. Roy. Soc. B *163B*, 224 (1965)
207. Pirt, S. J.: Principles of microbe and cell cultivation; Oxford: Blackwell 1975
208. Davis, P. E., Cohen, D. L., Whitaker, A.: Antonie v. Leeuwenhoek *46*, 391 (1980)
209. Stouthamer, A. H., Bettenhaussen, C.: Biochim. Biophys. Acta *301*, 53 (1973)
210. Esener, A. A., Roels, J. A., Kossen, N. W. F.: Biotechnol. Letters *3(1)*, 15 (1981)
211. Forrest, W. W., Walker, D. J.: Adv. Microb. Physiol. *5*, 213 (1971)
212. Sinclair, C. G., Topiwala, H. H.: Biotech. Bioeng. *12*, 1069 (1970)
213. Stouthamer, A. H.: Methods Microbiol. *1*, 629 (1969)
214. Stouthamer, A. H.: Yield studies in microorganisms, in: Patterns of Progress: (3)1 1976
215. Stouthamer, A. H.: Energetic aspects of the growth of micro-organisms, in: Symp. Soc. Gen. Microbiol. (28); Microbial energetics, p. 285, (Haddock, B. A., Hamilton, W. A., eds.) Cambridge: Univ. Press 1977
216. Stouthamer, A. H., Bettenhaussen, C.: Arch. Microbiol. *102*, 187 (1975)
217. Tempest, D. W.: TIBS *3*, 180 (1978)
218. Kuenen, J. G.: Arch. Microbiol. *122*, 183 (1979)
219. Oshima, T., Imahori, K.: J. Biochem. *75*, 179 (1974)
220. Han, M. H.: J. Theor. Biol. *35*, 543 (1972)
221. Ratkowsky, D. A., Olley, J., McMeekin, T. A., Ball, A.: J. Bacteriol. *149(1)*, 1 (1982)
222. Coultate, T. P., Sundaram, T. K.: ibid. *121*, 55 (1975)
223. Mohr, P. W., Krawiec, S.: J. Gen. Microbiol. *121*, 311 (1980)
224. Ng, T. K., Zeikus, J. G.: J. Bacteriol. *150(3)*, 1391 (1982)
225. Lamed, R., Zeikus, J. G.: ibid. *141(3)*, 1251 (1980)
226. Rogers, P. L., Lee, K. J., Skotnicki, M. L., Tribe, D. E.: Adv. Biochem. Eng. *23*, 37 (1982)
227. Cysewski, G. R., Wilke, C. R.: Biotech. Bioeng. *20*, 1421 (1978)
228. Andreesen, J. R., Schaupp, A., Neurauter, C., Brown, A., Ljungdahl, L. G.: J. Bacteriol. *114(2)*, 743 (1973)
229. Ben-Bassat, A., Lamed, R., Zeikus, J. G.: ibid. *146(1)*, 192 (1981)

230. Daniels, L., Fuchs, G., Thauer, R. K., Zeikus, J. G.: ibid. *132(1)*, 118 (1977)
231. Aunstrup, K., Andresen, O., Falch, E. A., Nielsen, T. K.: Production of microbial enzymes, in: Microbial Technology. Microbial Processes. I, p. 281, (Peppler, H. J., Perlman, D., eds.), New York: Acad. Press 1979[2]
232. Erickson, R. J.: Industrial application of *Bacilli*: A review and prospectus, in: Microbiology — 1976, p. 406, (Schlessinger, D., ed.) ASM Publ. 1976
233. Kula, M. R.: Chemie in unserer Zeit *2*, 61 (1980)
234. Freese, E., Fujita, Y.: Control of enzyme synthesis during growth and sporulation, in: Microbiology — 1976, p. 164, (Schlessinger, D., ed.) ASM Publ. 1976
235. Priest, F. G.: Bacteriol. Rev. *41*, 711 (1977)
236. Schaeffer, P.: ibid. *33(1)*, 48 (1969)
237. Welker, N. E., Campbell, L. L.: J. Bacteriol. *86*, 681 (1963)
238. Welker, N. E., Campbell, L. L.: ibid. *86*, 687 (1963)
239. Keay, L., Moseley, M. H., Anderson, R. G., O'Connor, R. J., Wildi, B. S.: Biotech. Bioeng. Symp. *3*, 63 (1972)
240. Sidler, W.: Versuche zur Temperaturadaptation thermophiler Bacillen und Produktion, Isolation und Charakterisierung extrazellulaerer neutraler Proteinasen mit verschiedener Thermostabilität aus *Bacillus stearothermophilus* and caldoaktiven Bacillen. PhD thesis, Swiss Federal Inst. of Technology, Nr. 5404 (1974)
241. Endo, S.: J. Ferment. Technol. *40*, 346 (1962)
242. Ohta, Y.: J. Biol. Chem. *242(3)*, 509 (1967)
243. Ohta, Y., Ogura, Y., Wada, A.: Biol. Chem. *241(24)*, 5919 (1966)
244. Keay, L., Wildi, B. S.: Biotech. Bioeng. *12*, 179 (1970)
245. Levy, P. L., Pangburn, M. K., Burstein, Y., Ericsson, L. H., Neurath, H., Walsh, K. A.: Proc. Nat. Acad. Sci. USA *72(11)*, 4341 (1975)
246. Feder, J., Kuo, M. J., Wildi, B. S.: Develop. Ind. Microbiol. *18*, 267 (1977)
247. Sidler, W., Zuber, H.: FEBS Letters *25(2)*, 292 (1972)
248. Sidler, W., Zuber, H.: Experientia *33*, 800 (1977)
249. Stahl, S., Ljunger, C.: FEBS Letters *63*, 184 (1976)
250. Aunstrup, K.: Appl. Biochem. Bioeng. *2*, 27 (1979)
251. Maentsaelae, P., Zalkin, H.: J. Bacteriol. *141(2)*, 493 (1980)
252. Sidler, W., Zuber, H.: Eur. J. Appl. Microbiol. Biotechnol. *10*, 197 (1980)
253. Roncari, G., Zuber, H.: Protein Res. *1(1)*, 45 (1969)
254. Sidler, W., Zuber, H.: Eur. J. Appl. Microbiol. *4*, 255 (1977)
255. Klingenberg, P., Zickler, F., Leuchtenberger, A., Ruttloff, H.: Z. Allg. Mikrobiol. *19(1)*, 17 (1979)
256. Leuchtenberger, A., Klingenberg, P., Ruttloff, H.: ibid. *19(1)*, 27 (1979)
257. Taeufel, A., Behnke, U., Ruttloff, H.: ibid. *19(2)*, 129 (1979)
258. Ingle, M. B., Boyer, E. W.: Production of industrial enzymes by *Bacillus* species, in: Microbiology-1976, p. 420, (Schlessinger, D., ed.) ASM Publ. 1976
259. Grootegoed, J. A., Lauwers, A. M., Heinen, W.: Arch. Microbiol. *90*, 223 (1973)
260. Voordouw, G., Roche, R. S.: Biochem. *14(21)*, 4659 (1975)
261. Voordouw, G., Roche, R. S.: ibid. *14*, 4667 (1975)
262. Pfueller, S. L., Elliott, W. H.: J. Biol. Chem. *244(1)*, 48 (1969)
263. Sekiguchi, J., Okada, H.: J. Ferm. Technol. *50*, 801 (1972)
264. Welker, N. E., Campbell, L. L.: J. Bacteriol. *86*, 1196 (1963)
265. Pazlarova, Y., Fencl, Z., Tsaplina, I. A., Egorova, L. A., Loginova, L. G.: Mikrobiologiya *46(3)*, 450 (1977)
266. Ogasahara, K., Imanishi, A., Isezuma, T.: J. Biochem. *67(1)*, 65 (1970)
267. Ogasahara, K., Imanishi, A., Isezuma, T.: ibid. *67(1)*, 77 (1970)
268. Ogasahara, K., Yutani, K., Imanishi, A., Isezuma, T.: ibid. *67(1)*, 83 (1970)
269. Heinen, W., Lauwers, A. M.: Arch. Microbiol. *106*, 201 (1975)
270. Hasegawa, A., Imahori, K.: J. Biochem. *79*, 469 (1976)
271. Hasegawa, A., Miwa, N., Oshima, T., Imahori, K.: ibid. *79*, 35 (1976)
272. Shinomiya, S., Yamane, K., Oshima, T.: Biochem. Biophys. Res. Comm. *96(1)*, 175 (1980)
273. Campbell, L. L.: Arch. Biochem. Biophys. *54*, 154 (1955)
274. Yutani, K., Sasaki, I., Ogasahara, K.: J. Biochem. *74*, 573 (1973)
275. Hartmann, P. A., Tetrault, P. A.: Appl. Microbiol. *3*, 11 (1955)

276. Hartman, P. A., Wellerson, R., Tetrault, P. A.: ibid. *3*, 7 (1955)
277. Krishnan, T., Chandra, A. K.: Appl. Environm. Microbiol. *44(2)*, 270 (1982)
278. Boyer, E. W., Ingle, M. B., Mercer, G. D.: Starch *31*, 166 (1979)
279. Mercer, G. D., Boyer, E. W., Ingle, M. B.: Abstr. Ann. Meet. ASM *76*, 186 (1976)
280. Chiang, J. P., Alter, J. E., Sternberg, M.: Starch *31*, 86 (1979)
281. Buonocore, V., Caporale, C., deRosa, M., Gambacorta, A.: J. Bacteriol. *128(2)*, 515 (1976)
282. Uchino, F.: Agric. Biol. Chem. *46(1)*, 7 (1982)
283. Coleman, G., Elliott, W. H.: Biochem. J. *83*, 256 (1962)
284. Mannig, G. B., Campbell, L. L.: J. Biol. Chem. *236(11)*, 2952 (1961)
285. Heineken, F. G., O'Connor, R. J.: J. Gen. Microbiol. *73*, 35 (1972)
286. Garcia-Martinez, D. V., Shinmyo, A., Madia, A., Demain, A. L.: Eur. J. Appl. Microbiol. Biotechnol. *9(3)*, 189 (1980)
287. Johnson, E. A., Sakajoh, M., Halliwell, G., Madia, A., Demain, A. L.: Appl. Environm. Microbiol. *43(5)*, 1125 (1982)
288. Petre, J., Longin, R., Millet, J.: Biochimie *63*, 629 (1981)
289. Ng, T. K., Zeikus, J. G.: Biochem. J. *199*, 341 (1981)
290. Chen, W. P.: Proc. Biochem. *15*, 30 (1980)
291. Fogarty, W. M., Griffin, P. J., Joyce, A. M.: ibid. *9(6)*, 11 and *9(7)*, 27 (1974)
292. Yoshimura, S., Danno, G. I., Natake, M.: Biol. Chem. *30(10)*, 1015 (1966)
293. Lamed, R. J., Zeikus, J. G.: Biochem. J. *195*, 183 (1981)
294. Bader, J., Giesel, H., Simon, H.: Die Aktivitäten von Hydrogenase sowie Enoatreduktase und ihre Wechselbeziehungen beim Wachstum einiger Clostridien, in: Technische Mikrobiologie, 4. Symp. Berlin, p. 251, (Dellweg, H. ed.) 1979
295. Egerer, P., Simon, H.: Biotechnol. Letters *4(8)*, 501 (1982)
296. Krasna, A. I.: Enzyme Microb. Technol. *1*, 165 (1979)
297. Miyamoto, K., Hallenbeck, P. C., Benemann, J. R.: Appl. Environm. Microbiol. *38(3)*, 440 (1979)
298. Simon, H., Bader, J., Rambeck, B., Krezdorn, E., Tischer, W.: Stereospezifische Hydrierung mit Mikroorganismen und immobilisierten Enzymsystemen, in: Technische Mikrobiologie, 4. Symp. Berlin, p. 325, (Dellweg, H., ed.) 1979
299. Wong, C. H., Daniels, L., Orme-Johnson, W. H., Whitesides, G. M.: J. Am. Chem. Soc. *103*, 6227 (1981)
300. Drake, H.: J. Bacteriol. *150(2)*, 702 (1982)
301. Aragno, M.: FEMS Microbiol. Letters *3*, 13 (1978)
302. Graf, G., Thauer, R. K.: FEBS Letters *136(1)*, 165 (1981)
303. Suzuki, Y., Nakamura, N., Kishigami, T., Abe, S.: J. Biochem. *87*, 745 (1980)
304. Suzuki, Y., Imai, T.: Biochim. Biophys. Acta *705*, 124 (1982)
305. Miwa, N., Nakajima, H.: Measuring composition containing enzymes. Europ. Pat. Applic. # 81302249.8, Publ. # 0 043 181 (1982)
306. deRosa, M., Gambacorta, A., Sodano, G., Trabucco, A.: Experientia *37(6)*, 541 (1981)
307. Buonocore, V., Sgambati, 0., deRosa, M., Esposito, E., Gambacorta, A.: J. Appl. Biochem. *2*, 390 (1980)
308. deRosa, M., Gambacorta, A., Lama, L., Nicolaus, B., Buonocore, V.: Biotechnol. Letters *3(4)*, 183 (1981)
309. Drioli, E., Iori, G., Molinari, R., deRosa, M., Gambacorta, A., Esposito, E.: Biotech. Bioeng. *23*, 221 (1981)
310. Drioli, E., Iorio, G., Santoro, R., deRosa, M., Gambacorta, A., Nicolaus, B.: J. Mol. Catal. *14*, 247 (1982)
311. Guy, G. R., Daniel, R. M.: Biochem. J. *203*, 787 (1982)
312. Guy, G. R., Morgan, H. W., McIver, R. D.: Stereospezifische D- und L-Asparaginase sowie Verfahren zu ihrer Herstellung. Deutsche Offenlegungsschrift DE 31 15 809: 20 pp (1982)
313. McClelland, M.: Nucl. Acids Res. *9(24)*, 6795 (1981)
314. Vasquez, C., Venegas, A., Vicuna, R.: Biochem. Int. *3(3)*, 291 (1981)
315. Balch, W. E., Fox, G. E., Magrum, L. J., Woese, C. R., Wolfe, R. S.: Microbiol. Rev. *43(2)*, 260 (1979)
316. Wolfe, R. S., Higgins, I. J.: Int. Rev. Biochem., Microbial Chemistry *21*, 267 (1979)
317. Loll, U.: Neue Verfahren zur biologischen Abwasserreinigung, in: Biotechnologie. Forschung aktuell, p. 180, (BMFT, ed.), Frankfurt: Umschau Vlg. 1979

318. Strauch, D., Boehm, R.: Forum Mikrobiol. *3*, 121 (1979)
319. Jewell, W. J., Kabrick, R. M., Spada, J. A., Autoheated, aerobic thermophilic digestion with air aeration. US environmental protection agency report: EPA-600/S2-82-023 Project summary 1982
320. Hess, E., Breer, C.: Zbl. Bakt. Hyg. I. Abt. Orig. B *161*, 54 (1975)
321. Hess, E., Lott, G., Breer, C.: ibid. *158*, 446 (1974)
322. Loll, U.: Korrespondenz Abwasser *21(6)*, 135 (1974)
323. von Steldern, D., Ottow, J. C. G., Loll, U.: Z. Allg. Mikrobiol. *14(3)*, 229 (1974)
324. Loll, U., Ottow, J. C. G.: 'gwf'-wasser/abwasser *115(11)*, 511 (1974)
325. Loll, U.: Betriebswerte und Wirtschaftlichkeit der aerob-thermophilen Stabilisation. BMFT-Statusseminar, Juelich, 247 (1981)
326. Loll, U.: Kommunalwirtschaft *9-1977*, 1 (1977)
327. Kandler, O.: Stand und Möglichkeiten der Abwasserreinigung mit Hilfe thermophiler Mikroorganismen; BMFT-Studie # 01 VY 016-AA/NT/BCT 0127
328. Scharer, J. M., Moo-Young, M.: Adv. Biochem. Eng. *11*, 49 (1979)
329. Mackie, R. I., Bryant, M. P.: Appl. Environm. Microbiol. *41(6)*, 1363 (1981)
330. Temper, U., Steiner, A., Winter, J., Kandler, O.: Thermophile Methangärung — Stand und Aussichten, BMFT-Statusseminar, Jülich, 19 (1981)
331. Cooney, C. L., Ackerman, R. A.: Eur. J. Appl. Microbiol. *2*, 65 (1975)
332. Pfeffer, J. T.: Biotech. Bioeng. *16*, 771 (1974)
333. van den Berg, L., Kennedy, K. J.: Biotech. Letters *3(4)*, 165 (1981)
334. Kennedy, K. J., van den Berg, L.: ibid. *4(3)*, 171 (1982)
335. Loll, U.: Verfahrenstechnischer Entwicklungsstand der anaeroben Stabilisation von Klärschlämmen, in: Technische Mikrobiologie, 4. Symp. Berlin, p. 179, (Dellweg, H., ed.) 1979
336. Cooney, C. L., Snedecor, B. R., Levine, D. W., Ackerman, R. A., Lee, J.: Develop. Industr. Microbiol. *18*, 255 (1971)
337. Redl, B., Tiefenbrunner, F.: Eur. J. Appl. Microbiol. Biotechnol. *12*, 234 (1981)
338. Kabrick, R. M., Jewell, W. J., Salotto, B. V., Berman, D.: Inactivation of viruses, pathogenic bacteria and parasites in the autoheated aerobic thermophilic digestion of sewage sludges, in: Proc. of the 34th Industrial Waste Conf. Purdue Univ. Lafayette, Indiana. Ann Arbor Science, 771 (1979)
339. Goddard, M. R., Bates, J., Butler, M.: Appl. Environm. Microbiol. *42(6)*, 1023 (1981)
340. Brock, T. D., Rose, A. H.: Methods Microbiol. *3B*, 161 (1969)
341. Lauwers, A. M., Heinen, W.: Arch. Microbiol. *95*, 67 (1974)
342. Stolp, H., Starr, M. P.: Principles of isolation, cultivation, and conservation of bacteria, in: The prokaryotes, p. 135, (Starr, M. P., Stolp, H., Trueper, H. G., Balows, A., Schlegel, H. G., eds.), Berlin, Heidelberg, New York: Springer 1981
343. Sonnleitner, B., Cometta, S., Fiechter, A.: Biotech. Bioeng. *24*, 2597 (1982)
344. deRosa, M., Gambacorta, A., Bu'lock, J. D.: J. Bacteriol. *117(1)*, 212 (1974)
345. Cooney, C. L., Wang, D. I. C., Mateles, R. I.: Biotech. Bioeng. *11*, 269 (1968)
346. Aiba, S., Humphrey, A. E., Millis, N. F.: Biochemical Engineering, New York: Acad. Press 1973^2
347. Goto, E., Kodama, T., Minoda, Y.: Agric. Biol. Chem. *41(4)*, 685 (1977)
348. Kodama, T., Goto, E., Minoda, Y.: ibid. *40(12)*, 2373 (1976)
349. Sakaki, Y., Oshima, T.: J. Virol. *15(6)*, 1449 (1975)
350. Zeikus, J. G., Taylor, M. W., Brock, T. D.: Biochim. Biophys. Acta *204*, 512 (1970)
351. Gomez, R. F., Hernandez, P.: Glucose utilization by *Clostridium thermocellum*, in: Advances in Biotechnology, II; Fuels, chemicals, foods and waste treatment, p. 131, (Moo-Young, M., ed.), Oxford: Pergamon Press 1981
352. Wang, D. I. C., Fleischhaker, R. J., Wang, G. Y.: AIChE Sympos. Ser. *181(74)*, 105 (1978)
353. Shinomiya, S., Oshima, T., Yamane, K.: Agric. Biol. Chem. *46(2)*, 345 (1982)
354. Tanaka, T., Kawano, N., Oshima, T.: Biochem. *89*, 677 (1981)
355. Nagahari, K., Koshikawa, T., Sakaguchi, K.: Gene *10*, 137 (1980)
356. Hishinuma, F., Tanaka, T., Sakaguchi, K.: J. Gen. Microbiol. *104*, 193 (1978)
357. Farrell, J., Campbell, L. L.: Adv. Microb. Physiol. *3*, 83 (1969)
358. Neilson, N. E., MacQuillan, M. F., Campbell, J. J. R.: Can. J. Microbiol. *5*, 293 (1959)
359. Epstein, I., Grossowicz, N.: J. Bacteriol. *99*, 414 (1969)
360. Atkinson, A., Evans, C. G. T., Yeo, R. G.: J. Appl. Bacteriol. *38*, 301 (1975)

361. Bubela, B.: Austral. J. Biol. Sci. *21*, 439 (1968)
362. Buswell, J. A., Twomey, D. G.: J. Gen. Microbiol. *87*, 377 (1975)
363. Pozmogova, I. N.: Mikrobiologiya *44(3)*, 492 (1975)
364. Golovacheva, R. S., Loginova, L. G., Salikhov, T. A., Kolesnikov, A. A., Zaitseva, G. N.: ibid. *44(2)*, 265 (1975)
365. Suzuki, Y., Kishigami, T., Abe, S.: Appl. Environm. Microbiol *31(6)*, 807 (1976)
366. deRosa, M., Gambacorta, A., Bu'lock, J. D.: J. Gen. Microbiol. *86*, 156 (1975)
367. deRosa, M., deRosa, S., Gambacorta, A., Carteni-Farina, M., Zappia, V.: Biochem. J. *176*, 1 (1978)
368. deRosa, M., Gambacorta, A., Nicolaus, B., Bu'lock, J. D.: Phytochemistry *19*, 821 (1980)
369. Wiegel, J., Braun, M., Gottschalk, G.: Curr. Microbiol. *5*, 255 (1981)
370. Gordon, J., Jiminez, M., Cooney, C. L., Wang, D. I. C.: AIChE Sympos. Ser. *181(74)*, 91 (1978)
371. Nazina, T. N., Rozanova, E. P.: Mikrobiologiya *47(1)*, 142 (1978)
372. Eyzaguirre, J., Jansen, K., Fuchs, G.: Arch. Microbiol. *132*, 67 (1982)
373. Zinder, S., Mah, R. A.: Appl. Environm. Microbiol. *38(5)*, 996 (1979)
374. Brock, T. D., Brock, K. M., Belly, R. T., Weiss, R. L.: Arch. Microbiol. *84*, 54 (1972)
375. Furuya, T., Nagumo, T., Itoh, T., Kaneko, H.: Agric. Biol. Chem. *41(9)*, 1607 (1977)
376. Furuya, T., Nagumo, T., Itoh, T., Kaneko, H.: ibid. *44(3)*, 517 (1980)
377. Caldwell, D. E., Caldwell, S. J., Laycock, J. P.: Can. J. Microbiol. *22*, 1509 (1976)
378. Caldwell, D. E., Laycock, J. P.: Abstr. Ann. Meet. ASM *76*, 173 (1976)
379. Brock, T. D.: Symp. Soc. Gen. Microbiol. *19*, 15 (1969)
380. Loginova, L. G., Khraptsova, G. I., Egorova, L. A., Bogdanova, T. I.: Mikrobiologiya *47(5)*, 947 (1978)
381. Landolt-Boernstein: Eigenschaften der Materie in ihren Aggregatzuständen. Zahlenwerte und Funktionen aus Physik, Chemie, Astronomie, Geophysik und Technik; Bd. 2, (Bartels, J., Borchers, H., Hausen, H., Hellwege, K. H., Schaefer, K. L., Schmidt, E., eds.), Berlin, Göttingen, Heidelberg: Springer 1962
382. Kuhn, H. J.: Einfluss der Temperatur auf das Wachstum von Bacillus caldotenax. PhD thesis; Swiss Federal Inst. of Technology, Nr. 6435 (1979)
383. Stetter, K. O.: Nature *300*, 258 (1982)
384. Zeikus, J. G., Lamed, R. J.: US Patent # 4,352,885 (1982)
385. Cornet, P., Tronik, D., Millet, J., Aubert, J.-P.: FEMS Microbiol. Lett. *16*, 137 (1983)
386. Brannan, D. K., Caldwell, D. E.: Appl. Environm. Microbiol. *45(1)*, 169 (1983)
387. Hansson, G.: Biotech. Letters *4(12)*, 789 (1982)
388. Kargi, F., Robinson, J. M.: Biotech. Bioeng. *24*, 2115 (1982)
389. Kargi, F., Robinson, J. M.: Appl. Environm. Microbiol. *44(4)*, 878 (1982)

Production of Molecular Hydrogen in Microorganisms

Elena N. Kondratieva
Department of Microbiology, Moscow State University, Moscow V-234, USSR
Ivan N. Gogotov
Institute of Photosynthesis Academy of Sciences of the USSR,
142292 Pushchino, USSR

This review surveys recent data on microorganisms able to evolve molecular hydrogen, conditions of H_2 production and properties of enzymes catalyzing this process. The data are discussed with regard to the problems of renewable energy resources.

1 Introduction

Microorganisms able to produce molecular hydrogen include chemotrophs, i.e. organisms obtaining energy by oxidation of chemical substrates, and phototrophs, i.e. those using light as the source of energy. The ability is distributed both among prokaryotes and eukaryotes.

H_2 production in microorganisms is often of great importance for their metabolism and for their relation to other species in nature. Some hydrogen-producing microorganisms are also able to consume H_2.

In recent years special attention has been paid to H_2-producing microorganisms with regard to renewable resources of energy. This is explained by molecular hydrogen which is an efficient fuel used also in some branches of industry. However the production of hydrogen is still expensive today [1-10].

Some H_2-evolving microorganisms in mixed cultures are involved in the production of gaseous fuels such as methane [4,5,11-16].

Various H_2-forming microorganisms may be the sources of other useful products including some enzymes. Some of these enzymes may be used for the production and utilization of molecular hydrogen [10,17-21].

The ability of certain H_2-evolving microorganisms to fix molecular nitrogen is also of considerable importance [21-25].

Thus, the range of problems whose solution is connected with the study of H_2-producing microorganisms is very large.

Principal data concerning H_2-producing microorganisms are summarized in several reviews [10,21,26-37]. Some of these articles are dedicated only to H_2-producing chemotrophs [38-41], or phototrophs [42-55]. Furthermore, on some symposia and conferences the practical use of H_2-producing microorganisms was discussed [56-64,225,286].

In this paper recent studies on H_2-producing microorganisms and enzymes participating in the metabolism of molecular hydrogen are reported.

2 Chemotrophs

Most H_2-producing chemotrophic microorganisms belong to obligate and facultative anaerobes (Table 1). The strict anaerobes are particularly numerous. Some of them were discovered quite recently. They include such thermophiles as *Thermoanaerobium brockii* [65], *Thermoanaerobacter ethanolicus* [66], *Thermobacteroides acetoethylicus* [67] and some other species [68,69,521].

Among facultative anaerobes the enterobacteria and related microorganisms predominate. H_2 is also produced by *Bacillus* spp. [35,70,71], *Campylobacter* sp. [72] and *Alcaligenes eutrophus* [73].

Recently, some obligate aerobes were shown to be able to evolve H_2 [74-85]. They chiefly occur as species able to assimilate molecular nitrogen: *Rhizobium spp.* [75-81], *Azotobacter* spp. [82,83], *Azospirillum brasilense* [84,85] and some others [18,21,35,74].

Many chemotrophic H_2-producing bacteria are widely distributed in nature. Some of them are symbionts of humans and animals [21,30,71]. Most characteristic in this respect are the microorganisms of the rumen. They include several H_2-producing bacteria: *Ruminococcus albus*, *Rum. flavefaciens*, *Selenomonas ruminatium*, *Megasphae-*

Table 1. H_2-producing chemotrophic bacteria

Genera	Species	Ref.
	Strict anaerobes	
Acetobacterium	*Ac. woodii*	69)
Bacteroides	*B. biacutus*, *B. clostridiformis*, *B. fragilis*, *B. termitidis*	35)

Table 1. (continued)

Genera	Species	Ref.
Butyrivibrio	*But. fibrisolvens*	87)
Clostridium	*C. aceticum, C. acetobutylicum, C. butyricum, C. butylicum, C. botulinum, C. cellobioparum, C. fallax, C. felsineum, C. glycolicum, C. histoliticum, C. kluyveri, C. pasteurianum, C. perfringens, C. rubrum, C. sphenoides, C. sporogenes, C. sartagoformum, C. thermocellum, C. thermohydrosulfuricum*	19, 31, 142, 230, 336, 337, 340)
Desulfovibrio	*D. africanas, D. desulfuricans, D. gigas, D. vulgaris*	31, 96, 151)
Eubacterium	*Eub. cellulosolfens, Eub. limosum*	86)
Megasphaera	*M. elsdenii*	35)
Methanobacterium	*Met. ruminantium*	35)
Methanothrix	*Methanothrix* sp.	341)
Methanococcus	*Mc. vannielii*	11)
Mycoplasma	*Mycoplasma* sp.	339)
Peptococcus	*P. activus, P. aerogenes, P. anaerobius*	38)
Peptostreptococcus	*Pep. anaerobius*	35)
Ruminococcus	*Rum. albus, Rum. flavefaciens*	87, 86)
Sarcina	*Sar. maxima, Sar. ventriculi*	230)
Selenomonas	*S. ruminantium*	87)
Spirochaeta	*Sp. litoralis, Sp. stenostrepta, Sp. zuelzerae*	338)
Syntrophobacter	*Sb. wolinii*	68)
Syntrophomonas	*Syn. wolfei*	521)
Thermoanaerobium	*T. brockii*	63)
Thermoanaerobacter	*Tm. ethanolicus*	66)
Thermobacteroides	*Tb. acetoethylicus*	67)
Veillonella	*V. alcalescens, V. parvus*	27, 39)
	Facultative anaerobes	
Aeromonas	*Ar. hydrophila, Ar. punctata, Ar. salmonicida*	354)
Alcaligenes	*A. eutrophus*	73)
Bacillus	*Bac. macerans, Bac. polymyxa*	38, 230)
Campylobacter	*Campylobacter* sp.	72)
Citrobacter	*Cit. freundii*	109)
Escherichia	*E. coli*	342)
Enterobacter	*En. aerogenes*	29)
Hafnia	*H. alvei*	354)
Klebsiella	*K. Pneumoniae*	343)
Photobacterium	*Ph. phosphoreum*	38)
Proteus	*Pr. mirabilis*	31)
Serratia	*Ser. marcescens*	39)
Streptococcus	*St. paracitrovorus*	39)
	Aerobes and microaerophils	
Azotobacter	*Ab. chroococcum, Ab. vinelandii*	82, 83)
Azospirillum	*Az. braziliense*	84)
Azomonas	*Azomonas* sp.	74)
Methylomonas	*Met. albus*	74)
Mycobacterium	*Mc. flavum*	83)
Pseudomonas	*Pseudomonas* sp.	35)
Rhizobium	*Rz. leguminosarum, Rz. japonicum, Rz. trifolii*	76, 77, 79)
Xanthobacter	*X.* (= Corynebacterium) *autotrophicus*	18)

ra elsdenii, and others [86,87]. Some species of H_2-producing microorganisms are pathogens of humans, animals or plants [70,71].

Trichomonades and other anaerobic protozoa are known to produce H_2. The majority of them live in the digestive or uro-genital tract of some animals. The best studied is the cattle parasite *Trichomonas foetus* [10,35,36,88,89].

There are also data about H_2-producing protozoa in rumen [86], for instance *Dastytrichia ruminantium* [90].

A few H_2-producing chemotrophs only grow in complex media [70,71]. Others may grow in rather simple media, containing different organic compounds. Certain species exhibit autotrophy and obtain energy as a result of H_2 oxidation. Such a capacity is characteristic of *A. eutrophus* [70,71,91], some strains of *Rhizobium* spp. [92,93] and methanogenic bacteria [11,15,94,95].

There are as well data about the growth under chemolithotrophic conditions in the presence of H_2 of some sulfate-reducing bacteria [96–100], *Acetobacterium woodii* [101], *Clostridium aceticum* [102,103] and *Escherichia coli* [104–106].

The capability of H_2 oxidation has been found also in several other hydrogen-producing chemotrophs [21,40,31,93,107–109].

Generally, obligate and facultative anaerobic bacteria as well as protozoa evolve molecular hydrogen as a result of degradation of organic substrates, i.e. anaerobic degradation of organic compounds resulting in ATP synthesis at substrate level. Proteins, carbohydrates, alcohols, organic acids, amino acids, purines, and pyrimidines are among the organic compounds capable of degradation with H_2 evolution. Also the formation of some practically important organic products including acetic acid, ethanol, butanol, acetone and others [19,38,39,41,71] may result.

At present, great attention is paid to the use of new anaerobic microorganisms for the production of ethanol, acetate and other practically valuable compounds. These microorganisms include thermophilic H_2-producing species [19,20,110].

The capability of various chemotrophs to use organic substrates and to produce H_2 as the result of their degradation is not the same [11,33,35,38–41,70,71,111–116]. However, many chemotrophs produce H_2 by the use of different carbohydrates [19,32,36,71,90,110,113,117].

Cellulose-utilizing microorganisms are of particular significance as this substrate is synthesized by plants in great amounts [118]. *Clostridium cellobioparum, S. ruminantium, Rum. flavefaciens* and some other anaerobic bacteria convert cellulose with the production of H_2 [11,70,87,110]. Many H_2-forming bacteria are able to use such widely distributed compounds as hemicellulose and starch [19,39,70,110,119]. The production of molecular hydrogen in some processes allows microorganisms to oxidize particular substrates more efficiently [10,13,40,41,113,114]. But not all substrate can be directly oxidized by microorganisms with the production of H_2. In many cases, degradation steps are involved which lead to the formation of intermediates that are oxidized with the evolution of molecular hydrogen.

Many microorganisms evolve H_2 as the result of anaerobic oxidation of pyruvate, an intermediate metabolite of carbohydrates and some other compounds. This is the pathway of H_2 production in many *Clostridium* spp. and several other anaerobes: *T. brockii* [19], *Peptococcus anaerobius, Eubacterium limosum, M. elsdenii, Sarcina maxima, Sar. ventriculi, Rum. albus, Veillonella alcalescens*, certain sulfate-reducing bacteria and spirochetes [31,35,40,41,120].

In most cases, the degradation of pyruvate with H_2 evolution is catalyzed by pyruvate-ferredoxin oxidoreductase. The acetyl-CoA produced can be transformed into acetyl-phosphate which is used by many anaerobes for the synthesis of ATP by substrate-level phosphorylation [40,41]:

$$\begin{array}{ccc} \text{pyruvate} + \text{Fd} + \text{CoA} \rightarrow & \text{acetyl-CoA} + CO_2 + & \text{FdH}_2 \\ & \downarrow{\scriptstyle +P_i} & \downarrow \\ & \text{acetyl-phosphate} & \text{Fd} + H_2 \\ & \downarrow{\scriptstyle +ADP} & \\ & \text{acetate} + \text{ATP} & \end{array}$$

In this process, H_2 production is accompanied by the recycling of ferredoxin as well as CoA which may participate again in the oxidation of pyruvate followed by ATP synthesis.

Besides bacteria, a few H_2-producing protozoa are able to oxidize pyruvate in the same way. In some of these eukaryotic microorganisms pyruvate oxidation occurs in microbodies called hydrogenosomes [32,36,88,89,90].

In *Veillonella alcalescens* ferredoxin seems to participate both in H_2 production and in the transformation of hypoxanthine to xanthine [121] and oxidation of acetaldehyde to acetate by bacterium forma S associated with methanogenic species [36,122,123].

In some microorganisms the oxidation of organic substrates with the production of H_2 is linked to other electron carriers (Sect. 4.1.3).

In several cases H_2 production from reduced NAD is involved. This was found in *Clostridium kluyveri* and later in other species of *Clostridium* (*C. butyricum*, *C. pasteurianum*, *C. sphenoides*, *C. glycoliticum*). *Rum. albus*, *P. anaerobius*, *S. ruminantium* may also produce H_2 from NADH [10,35,40,120,124]. Probably, some other bacteria possess the same capacity.

H_2 formation from NADH in *Clostridium* spp. and some other microorganisms is catalyzed by NADH: ferredoxin oxidoreductase [40,41]. H_2 formation from NADH also allows the microorganisms to recycle the electron carriers and to form compounds needed for ATP synthesis.

The growth of some strict anaerobes is completely dependent on the degradation of organic substrates involving formation of H_2. A strict dependence is observed in microorganisms degrading organic compounds with the participation of NAD or NADP-linked dehydrogenase and producing H_2 from NADH [13,14]. *C. kluyveri* [40,41,113] using ethanol is an example.

Several other organic compounds [11,13,19,41,81,114] including fatty acids [19,68,116] may be degraded by anaerobic microorganisms if they produce molecular hydrogen.

The members of the enterobacteriaceae and some other facultative anaerobes produce H_2 from formate, a metabolite itself [39,41,71,113]. Obligate anaerobes also include species capable to oxidize formate with the evolution of H_2. Such microorganisms were found among methanogens [10,94,125,126] and sulfate-reducing bacteria [28,31,35].

Some microorganisms can produce formate from formyltetrahydrofolate [40].

But in most cases formate is produced from pyruvate by pyruvate-formate lyase [41, 113, 127]:

pyruvate → acetyl-CoA + formate

The degradation of formate with the evolution of H_2 by *E. coli* and some other microorganisms is catalyzed by formate: hydrogen lyase:

formate → CO_2 + H_2

This enzyme complex includes formate dehydrogenase, hydrogenase and electron carriers. But its nature has not yet been elucidated [28, 41, 71, 113, 128, 129].

The production of H_2 from formate by *E. coli* and some other bacteria prevents acidification inhibiting the growth of microorganisms [32, 130].

There are different possibilities for some chemotrophic microorganisms of producing H_2. For example, *R. albus* may produce H_2 with the participation of pyruvate: ferredoxin oxidoreductase and NAD: ferredoxin oxydoreductase [10, 40, 41].

Sulfate-reducing bacteria and some other microorganisms are characterized by their different possibilities of producing H_2 [31, 35, 40, 41, 120]. But in all cases H_2 production is catalyzed by hydrogenases [10, 18, 21, 35–37].

In contrast, H_2 production in aerobic bacteria is usually catalyzed by nitrogenase, i.e. by the enzyme providing the capacity for N_2 assimilation [10, 21, 37, 52, 75, 76, 83, 131–133]. The facultative anaerobic bacterium *Klebsielle pneumoniae* is able to produce H_2 with the participation of both hydrogenase and nitrogenase [134]. There is also evidence for H_2 production with the participation of hydrogenase and nitrogenase in *Azospirillum brazilense* [84]. Probably, this ability is also displayed by other microorganisms.

Thus, H_2 production by chemotrophic microorganisms as well as the importance of this process are rather diverse. Moreover, H_2 evolution by microorganisms depends on several factors. Often, the nature of the oxidized substrate is important [21, 43]. For example, *V. alcalescens* produces H_2 from lactate and hypoxanthine whereas no H_2 is formed from succinate [43].

The production of molecular hydrogen by the cells of *E. coli* and some other bacteria is significantly increased in the presence of formate in the growth media. The production of H_2 from formate is increased also at low pH values of the medium [31, 32, 109, 130, 135].

The formation of H_2 by different microorganisms in media deficient in iron is sharply decreased or stopped. As a result, the ratio and composition of other products may change [31, 32, 35].

Most chemotrophs evolve H_2 under anaerobic conditions. Only in *Azotobacter*, in some other N_2-fixing bacteria and in root nodules the formation of H_2 depends on the presence of O_2. As already mentioned, in such microorganisms the production of H_2 is catalyzed by nitrogenase, the activity of the latter requiring energy. However, at high oxygen levels the production of hydrogen by aerobic microorganisms is usually inhibited [23, 83, 84].

There is as well evidence that *A. eutrophus* may produce H_2 under microaerobic conditions [73]. In general however, facultative anaerobic bacteria evolve H_2 in anaerobic cultures [31,32,35,109]. The cells of such microorganisms grown under aerobic conditions do not produce H_2 or evolve molecular hydrogen only in small amounts.

The contact of microorganisms grown under anaerobic conditions with O_2 usually results in the inhibition of H_2 production, though this inhibition may be reversible [31,109,135]. A stronger inhibitory effect of O_2 on hydrogen production is exerted in strict anaerobes. But immobilized cells of obligate anaerobe *C. butyricum* produce H_2 even under aerobic conditions [10,136–138].

The formation of H_2 in bacteria capable of both aerobic and anaerobic respiration depends on the presence of compounds used as electron acceptors. For example, the production of H_2 by *E. coli* [31] and *Citrobacter freundii* [109] is repressed by nitrate.

A reduction or total inhibition of H_2 production by some sulfate-reducing bacteria is observed in the presence of sulfate or fumarate [23,40,41,139]. Some oxidants may inhibit H_2 formation in bacteria providing energy only as a result of metabolism. This is explained by their capacity to use such compounds as accessory electron acceptors [41]. For instance, H_2 evolution in *T. brockii* from glucose and pyruvate is diminished in the presence of acetone which is reduced to isopropanol [19,140]. Molecular nitrogen inhibits H_2 production by N_2-fixing bacteria [21,75,83].

Finally, the accumulation of molecular hydrogen may inhibit its own formation in many microorganisms [21,32,140–145]. As a result, an increased formation of reduced organic products is usually observed [35,40,41,121,143–145].

In many cases, high partial hydrogen pressures inhibit the growth of microorganisms [69,141,142,144,146]. This is mainly observed when anaerobic degradation of organic compounds is obligatorily related to H_2 evolution.

The use of ethanol, fatty acids and some other substrates by *C. kluyveri* [40,41] and other anaerobic bacteria [13,103,121,146] may serve as examples. The inhibitory effect of molecular hydrogen is usually especially strong when H_2 formation is linked to the oxidation of NADH. The redox potential of NADH/NAD ($E_0' = -320$ mV) is more positive than that of H_2/H^+ ($E_0' = -420$ mV). For this reason, the oxidation of NADH to NAD and H_2 is possible only at a very low partial pressures of molecular hydrogen [40] (about 1.5×10^{-3} atm $= 10^{-6}$ M H_2 [40,41]). In nature the low partial pressure of H_2 is maintained as the result of interactions between H_2-producing and H_2-consuming microorganisms. Sometimes, it is difficult to isolate them even in pure cultures. A well-known example is "*Methanobacillus omelianskii*" considered for a long time as methanogenic bacterium. Now it is known that this is a mixture of two microorganisms. One of them (forma S) oxidizes ethanol and some other organic compounds with the evolution of H_2. The second one (*Methanobacterium bryantii*) uses H_2 and CO_2 for the growth and production of methane [122,123].

This type of interactions between microorganisms is very important in rumen. Up to 80 l of H_2 may be formed in rumen of a cow every day as a result of the fermentation of cellulose and the substrates of some other plants by different microorganisms. But there is no accumulation of molecular hydrogen because it is consumed by the methanogenic bacteria [87].

Associations of microorganisms including H_2-producing and H_2-consuming species are also distributed in other ecosystems [11,23–26,41,94,143]. *Ruminococcus* spp. [121], *Selenomonas ruminantium* [147], *C. cellobioparum*, *C. thermocellum* [141,148], *Cit. freun-*

dii[149)], *Ac. woodii*[150)], *T. brockii*[110)], and *S. wolfei*[521)] have been found to play the role of H_2-producing components of this type of associations.

Cases of concomitant development of methanogenic and sulfate-reducing bacteria able to produce H_2 from some substrates have also been discovered [11,16,114,146,151)].

But in most cases, the sulfate-reducing bacteria themselves consume H_2 evolved by other microorganisms [31,96–100)]. For instance, *Synthrophobacter wolinii* and *Syntrophomonas wolfei* use some fatty acids for growth and evolve H_2 in the presence of *Desulfovibrio* spp. which oxidize molecular hydrogen [68,521)].

The interspecies transfer of H_2 also takes place between other microorganisms. For instance, *Rum. albus* degrades glucose with the production of H_2, and *Vibrio succinogenes* consumes hydrogen for the reduction of fumarate to succinate. As a result, the growth of both bacteria is stimulated, and *Rum. albus* produces more H_2 [36,40,41,143)].

In addition to anaerobes there are many aerobic microorganisms capable of consuming molecular hydrogen [21,30,91,152)].

Thus, there is evidence that H_2 belongs to metabolites coupling in nature the activity of several microorganisms and first of all of many chemotrophic anaerobes. Only associations of microorganisms including H_2-producing species seem to be able to degrade several organic compounds under anaerobic conditions [13–16,153–155)].

The maximum amount of H_2 may be formed by *Rum. albus* and some other bacteria degrading carbohydrates with the production of acetate. These acetogenic bacteria can produce up to 4 mol of H_2 per 1 mol of decomposed sugar:

$$4 \text{ glucose} \rightarrow 2 \text{ acetate} + 2\, CO_2 + 4\, H_2$$

But this quantity of hydrogen may be evolved only at a low p_{H_2}. At $p_{H_2} = 1.0$ atm H_2 production by *Rum. albus* does not exceed 2.0–2.6 mol per mol glucose [40,41,143,156)].

Several *Clostridium* spp., *E. coli* and some other chemotrophs also belong to the active H_2-producing microorganisms (Table 2). *C. butyricum* may evolve up to 2 mol of H_2 per 1 mol of decomposed sugar:

$$2 \text{ glucose} \rightarrow \text{acetate} + \text{butyrate} + 4\, CO_2 + 4\, H_2$$

Some H_2-producing bacteria are able to grow on cheap plant substrates containing carbohydrates. *C. perfringens* strain C was proposed for the production of H_2 from carbohydrate substrates. This microorganism growing in a 10-l fermentor may form up to 23 l H_2 per hour. This is sufficient for a fuel cell operating with a power up to 3.5 W. Such a method of H_2 production may be recommended for some developing countries possessing great resources of raw plant materials [157)].

Immobilized cells of *Clostridium* spp. and some other bacteria may be also used for H_2 production [10,136–138)].

Among other microorganisms *E. coli* and *Cit. freundii* produce H_2 at a high rate and in rather high amounts. Metabolism of glucose and some other sugars by these bacteria is connected with the production of H_2 and CO_2 in the ratio of 2.3–3.0. This ratio is higher than in other bacteria, *Rum. albus* included [35)].

Table 2. Amounts of H_2 (μmoles per 100 μmoles of substrate) produced by several bacteria

Bacteria	Substrates	H_2	Ref.
Bacillus polymyxa	glucose, mannitol	70.9–170.0	[38]
Clostridium acetobutylicum	starch	135.0	[39]
C. butyricum	glucose	235.0	[19]
C. butylicum	starch	78.0	[39]
C. kluyveri	ethanol + acetate	1.1	[39]
C. perfringens	glucose	214.0	[39]
C. thermocellum	cellulose	122.0	[19]
C. thermohydrosulfuricum	starch	11.0	[39]
Escherichia coli	glucose	75.0	[38]
Eubacterium limosum	glucose	74.0	[38]
Peptococcus aerogenes	serine, threonine, adenine, uracil	49–96.0	[38]
Photobacterium phosphoreum	glucose	54.0	[38]
Ruminobacter flavefaciens	cellulose	33.0	[19]
Ruminococcus albus	carbohydrates	257.0	[19]
Sarcina maxima	carbohydrates	230.0	[19]
S. ventriculi	hexose	41.0	[19]
Thermoanaerobium brockii	hexose	50.0	[19]
Veillonella alcalescens	lactate	14.0	[39]

A patent is granted for the use of *E. coli* cells for H_2 production from formate [158], formed as a by-product of some industrial processes.

It is also possible to determine formate quantitatively by the use of immobilized cells of bacteria cleaving this substrate with the production of H_2 which is measured by an electrochemical method [137].

But calculations show that the efficiency of energy conversion to H_2 does not exceed 33% of combustible energy in organic substrates. In practice, H_2 yields are lower and the efficiency of energy conversion is approximately 20% [156].

Thus, since the production of hydrogen is limited it is not used on a larger scale for practical purposes. But if H_2 formation is coupled with the production of other important compounds or with the removal of wastes it may be useful in some cases. Production of acetone and butanol are examples [19].

But more important is the production of methane. Microorganisms involved in this process usually include H_2-producing species and bacteria which consume H_2 with formation of CH_4. Such mixed cultures of microorganisms grow on different substrates, including the biomass of higher plants and algae, as well as on agricultural and other kind of wastes [4–7,12,19,60,62]. When the organic substrate is decomposed with the reduction of methane the efficiency of energy conversion may reach 85% [156].

Great attention is paid to microbial production of the so-called biogas containing methane [4–7,12,13,21,60,62]. This method is used in many places [6,21,60]. It is particularly important as energy source in some developing countries [60].

Manure is considered to be a suitable raw material for microbial methane production in nations with high development of livestock raising [60,62]. The formation of methane by microorganisms may be coupled with the synthesis of other valuable compounds, for instance, of vitamin B_{12} [159]. Moreover, the use of mixed cultures of microorganisms producing CH_4 allows to keep the environment clean from some wastes [4,21,60].

Of practical importance is the production of H_2 by N_2-utilizing bacteria, especially by *Rhizobium* spp. whose activity provides soil enrichment by nitrogen compounds.

Some strains of *Rhizobium* spp. consume up to 25% and even more energy for H_2 evolution [75,76,133,160] rather than for N_2-fixation. This is not desirable for the practical use of N_2-fixing microorganisms in agriculture. Therefore, it is important to select strains *of Rhizobium* spp. and other N_2-fixing bacteria not releasing H_2 (Sect. 4.3).

3 Phototrophs

The phototrophic organisms producing H_2 include several bacteria and algae. The majority of them are aquatic species [21,35,42–55]. There are also data on H_2 evolution by higher plants [161,162]. But this problem requires further experiments.

3.1 Purple Bacteria

Many investigations of H_2 evolution by phototrophic microorganisms concern the purple bacteria. These microorganisms perform photosynthesis without O_2 evolution and use sulfide, thiosulfate, sulfur, organic compounds, or molecular hydrogen as electron donors. Purple bacteria are able to assimilate CO_2 and several organic substrates [163–166] as the source of carbon.

Some purple bacteria may grow in the dark under anaerobic and (or) aerobic conditions. Under such conditions, they absorb energy as a result of anaerobic or aerobic respiration using organic compounds [165–168].

Certain purple bacteria can grow in the dark under chemolithoautotrophic conditions and absorb energy due to the oxidation of inorganic sulfur compounds [165,169–171,523] or H_2 [172,173]. Most species are capable to fix N_2 [53,163,164,174,175].

It has been known for a long time that some purple bacteria may produce H_2 in the dark using endogenous or exogenous organic compounds such as sugars, glycerol, pyruvate, and others [27,42,176]. Some strains of *Rhodospirillum rubrum* evolve H_2 in the dark as a result of formate oxidation which is produced from fructose and pyruvate. The degradation of formate is catalyzed by pyruvate: formate lyase [127,130,171–180]. The same capacity was found in *Rhodopseudomonas palustris* [167,181,182]. *Chromatium* sp. and *Thiocapsa roseopersicina* may also produce H_2 in the dark but only from pyruvate [21,182,183].

There are strains of *Rhodopseudomonas gelatinosa* and *R. rubrum* known to be able to grow in the dark under anaerobic conditions using CO. The cells of *Rh. gelatinosa* show the capacity for such a reaction [167,184,185]:

$$CO + H_2O \rightarrow CO_2 + H_2$$

Probably, some chemotrophic anaerobic bacteria are also able to use CO with H_2 formation [186].

H_2 production by purple bacteria in the dark is usually catalyzed by hydrogenases [179,180,187]. However, in some cases nitrogenase may be active in this process [179].

Evidence of the production of H_2 by purple bacteria in the light is more extensive [8, 10, 21, 33, 35, 35, 43–45, 51–53, 176]. This process called H_2 photoproduction or photoevolution is characteristic of many purple bacteria (Table 3).

In the light H_2 production is possible by growing cultures [21, 42, 176, 133–190], cell suspensions [21, 43, 44, 53, 182, 187, 190–198] and immobilized cells [8]. The rate of hydrogen production by purple bacteria in the light is usually greater than in the dark. In both cases, this process takes place under anaerobic conditions. The cells of purple bacteria grown under conditions of strong aeration usually do not produce any H_2 [21, 53, 187].

Photoproduction of H_2 by purple bacteria may occur at low light intensity. But up to 4000–6500 lx the rate of this process increases. However, at very high light intensities the evolution of H_2 may proceed more slowly [194, 197, 198].

Photoevolution of H_2 by purple bacteria as well as CO_2 assimilation and growth occur both by irradiation with white light and infrared light up to wavelengths of 900 nm and for some species up to 1100 nm. No other phototrophs are able to use these wavelengths for growth and H_2 production [163–165].

Optimal temperatures for H_2 production by many purple bacteria are 30–40 °C. But for some species higher temperatures are required [21, 199].

H_2 evolution depends on the supply of growing cultures with iron [93], molybdenum [53, 200, 201] and manganese [202].

In contrast to H_2 production in the dark, photoevolution of hydrogen also depends

Table 3. H_2-producing phototrophic microorganisms

Genera	Species	Ref.
	Purple sulfur bacteria	
Chromatium	*Chr. minutissimum, Chr. vinosum*	217, 344)
Ectothiorhodospira	*Ec. mobilis, Ec. shaposhnikovii*	188)
Thiocapsa	*Th. roseopersicina*	182)
	Purple nonsulfur bacteria	
Rhodospirillum	*R. rubrum, R. tenue*	42)
Rhodomicrobium	*Rhodom. vannielii*	345)
Rhodopseudomonas	*Rh. acidophila, Rh. capsulata, Rh. gelatinosa, Rh. palustris, Rh. sphaeroides, Rh. sulfidophila, Rh. viridis*	42, 181, 182)
	Green sulfur bacteria	
Chlorobium	*Chl. limicola* forma thiosulfatophilum	346)
Pelodyction	*Pelodyction* sp.	45)
	Cyanobacteria	
Anabaena	*An. azollae, An. cylindrica, An. flos-aquae, An. variabilis*	244, 298, 347)
Calothrix	*Cal. membranacea, Cal. scapulorum*	350)
Chlorogloea	*Chlorogloea* sp.	235)
Gloeocapsa (= *Chroococcus*)	*Gloeocapsa* sp.	351)
Gloeothece	*Gloeothece* sp.	351)
Lyngbya	*Lyngbya* sp.	351)
Mastigocladus	*Mas. laminosus, Mas. thermophilus*	257, 258)
Nostoc	*Nostoc.* sp., *Nos. muscorum*	247)
Oscillatoria	*O. limnetica*	239)
Plectonema	*Plectonema* sp., *Pl. borianum*	298, 299)

Table 3 (continued)

Genera	Species	Ref.
Scytonema	*Scytonema* sp.	348)
Spirulina	*Sr. maxima, Sr. platensis*	18, 349)
Stygonema	*Stygonema* sp.	348)
Synechococcus	*Syn. elongatus, Synechococcus* sp. (= *Anacystis nidulans*)	348)
Tolypothrix	*Tolypothrix* sp.	348)
	Green algae	
Ankistrodesmus	*Ankistrodesmus* sp., *Ank. brauni, Ank. falcatus, Ank. stripitatum*	352, 353)
Chlamydomonas	*Chlamydomonas* sp., *Chl. debaryana, Chl. disosmos, Chl. eugametos, Chl. humicola, Chl. moewussii, Chl. reinhardi*	296, 297, 353, 356)
Chlorella	*Ch. autotrophica, Ch. fusca, Ch. homosphaera, Ch. kessleri, Ch. protothecoides, Ch. pyrenoidosa, Ch. sorociniana, Ch. vacuolata, Ch. vulgaris*	294, 297, 298, 355)
Codium	*Codium* sp., *Cod. fragilis*	56, 291)
Coelastrum	*Cl. proboscideum*	359)
Dunaliella	*Dunaliella* sp., *Dun. salina*	56)
Kirchneriella	*Kir. lunaris*	297)
Scenedesmus	*Sc. obliquus, Sc. quadricauda*	297, 357)
Selenastrum	*Selenastrum* sp., *Sel. gracile*	359)
	Diatom algae	
Nitzchia	*Nz. ovalis*	18)
	Red algae	
Ceramium	*Ceramium* sp., *Cer. rubrum*	291)
Chondrus	*Cd. crispys*	56)
Corallina	*Corallina* sp., *Cor. officinalis*	56, 291)
Callithamnion	*Callithamnion* sp.	56)
Porphyridium	*Pd. aeragineum, Pd. cruentum, Porphyridium* sp.	291, 360)

on the source of nitrogen. H_2 formation in the light occurs in the presence of amino acids, N_2 or in cells growing in the presence of low amounts of ammonium [21, 44, 51, 53, 174, 182, 191–195]. An increase in H_2 formation also occurs after cell incubation under nitrogen deficiency conditions [524].

Cells grown in the presence of high NH_4^+ concentrations do not evolve H_2 in the light. This process may begin after a lag-phase but the rate of H_2 production is low [42, 44, 176, 182, 187, 203, 204]. Similar effects on H_2 photoevolution are brought about by glutamine, asparagine, urea, and nitrates [21, 187, 464].

Ammonium causes a reversible inhibition of photoevolution of H_2 in different purple bacteria [53, 174, 182, 187, 194, 195, 203–208]. A slighter effect display glutamine, asparagine and in some cases nitrate [187]. Molecular nitrogen and acetylene also inhibit H_2 photoevolution by purple bacteria [8, 21, 53, 187]. But CO which inhibits hydrogen evolution catalyzed by hydrogenases in many microorganisms does not reveal such an effect on photoproduction of H_2 by purple bacteria [182, 187].

Mutants of *Rh. capsulata* [210] and *Rh. palustris* [201] were isolated which are devoid of the capacity for N_2 assimilation and photoevolution of H_2. On the other hand, there are mutants of *R. rubrum* [211], *Rh. capsulata* [212] and *Rh. sphaeroides* [213] derepressed by nitrogenase. These mutants are able to assimilate N_2 and produce H_2 independently of the presence of ammonium.

All this evidence indicates that H_2 photoproduction in purple bacteria is catalyzed by nitrogenase through the majority of strains possess hydrogenases too. In general, in the light, however, the hydrogenases of purple bacteria are involved in the consumption of H_2 [21,53,165,174,187].

Purple bacteria may produce certain amounts of H_2 in the light as well as in the dark from endogenous reserve compounds [42,44,214]. However, considerably higher amounts are produced in the presence of exogenous substrates. Such substrates may be different organic compounds used by many purple bacteria in photosynthesis as electron donors. They include some alcohols, sugars and organic acids [8,21,42–44,53,174,182,194,195,201].

However, there are different possibilities of using different organic compounds in various purple bacteria. Several species are not able at all to oxidize organic substrates with the production of H_2 [163–165]. At the same time, some purple bacteria may photoevolve H_2 as a result of oxidation of thiosulfate [174,182,204,215].

H_2 evolution by purple bacteria in the light proceeds with participation of their photosynthetic apparatus which supplies this process with energy and a reduced electron carrier. Most probably, this carrier is ferredoxin [21].

According to many data purple bacteria contain only cyclic electron transport systems connected with photoreaction centers. The reduction of NAD and ferredoxin occurs as a result of energy-dependent reversed electron flow from different exogenous H-donors. Only in a few cases are the electron carriers reduced without consumption of energy [21,43,166,216].

But according to other authors [8,217] purple bacteria contain cyclic and non-cyclic electron transport systems coupled with photoreaction centers.

The relation between nitrogenase activity in purple bacteria and their photosynthetic activity is not obligatory. Some of these microorganisms may assimilate N_2 in the dark [204,218,219]. It was also shown that they are able to produce H_2 in the dark with the participation both of hydrogenase and nitrogenase [179].

Compounds allowing an intensive and considerable H_2 photoproduction in several purple bacteria include malate, succinate, pyruvate, and especially lactate [8,21,42,44,53,174,182,191–194,202,221]. But some species and strains generate large amounts of H_2 in the presence of some other substrates. A mutant of *Rh. sphaeroides* is able to produce H_2 from gluconate and glucose [53]. Upon illumination the cells of this mutant may completely oxidize glucose to CO_2 and H_2:

$$C_6H_{12}O_6 \rightarrow 6\,CO_2 + 6\,H_2$$

Earlier, it was shown that cells of *R. rubrum* may oxidize acetate, malate, fumarate, and succinate in a similar way. This process is connected with the function of tricarboxylic acid cycle under light anaerobic conditions [43,53,224].

Some purple bacteria may completely oxidize formate [182] and/or lactate [8] to CO_2 and H_2:

$$C_3H_6O_3 + 3\,H_2O \rightarrow 2\,CO_2 + 6\,H_2$$

This capacity is of interest with a view to practical utilization of purple bacteria for H_2 production because these compounds are cheap and are present in some wastes.

The utilization of one kg of lactate by *Rh. capsulata* may give more than 1350 l of H_2 [8]. Cultures of *R. rubrum* growing on media with lactate-containing wastes produce H_2 for up to 80 days. The rate of this process was 20 ml H_2 h^{-1} g^{-1} dry biomass [190]. According to available calculations the use of some wasts by phototrophic bacteria may give up to 50 kg H_2 m^{-2} a year [8, 12]. The communication concerning the ability of some purple bacteria (probably in mixed cultures) to produce H_2 on cellulose media [8, 223] is also interesting.

Thus, considerable progress in studying the conditions of H_2 production by purple bacteria is being made today. Active strains of these bacteria have been selected [8, 12, 53, 194, 195, 215].

An especially high rate of H_2 production is revealed by *Rh. capsulata*, strain B10 (Table 4). This bacterium in a continuous culture on lactate medium containing small amounts of ammonium evolves up to 500 ml H_2 h^{-1} g^{-1} dry biomass.

The efficiency of light energy conversion with the production of H_2 by purple bacteria cells reaches 2.8% [8, 222]. According to data on light-conversion efficiency in different phototrophs this is not a small value [4, 5, 216, 217].

Table 4. Maximum rate of H_2 production by phototrophic microorganisms

Microorganisms	H_2 (h^{-1} g^{-1} dry biomass)		Ref.
	ml	mmol	
	Purple bacteria		
Rhodospirillum rubrum S1	146.0	6.5	8)
Rhodopseudomonas palustris A	54.0	2.4	201)
Rh. acidophila DSM 7050	22.7	1.0	21)
Rh. vannielii G	2.7	0.1	214)
Rh. sphaeroides 2R	29.8	1.3	21)
Rh. capsulata SL	130–150.0	5.8–6.7	194)
Rh. capsulata LB2	178.0	7.9	8)
Rh. capsulata B10	300.0–500.0	13.4	21)
Rh. capsulata SCJ	168.0	7.5	8)
Chromatium vinosum D	9.0	0.4	217)
Ectothiorhodospira shaposhnikovii 1K	3.8	0.2	198)
Thiocapsa roseopersicina BBS	13.4	0.6	182, 198)
	Cyanobacteria		
Anabaena cylindrica B629	4.9–40.0	0.2–2.0	244)
An. flos-aquae	4.9	0.2	2)
An. variabilis	32.0	1.3	8)
Spirulina platensis	9.0	0.4	21)
	Green algae		
Chlorella vulgaris K	4.0	0.2	300)
Ch. pyrenoidosa ICC 251	1.6	0.1	297)
Chlamydomonas reinhardii 137C	45.7	2.0	296)
Chl. dysosmos ICC 342	6.0	0.3	297)
Chl. moevussii ICC 97	8.2	0.4	297)
Scenedesmus obliquus D_3	6.0	0.3	297)

Certain purple bacteria as well as some chemotrophic microorganisms evolve H_2 at a high rate in the presence of some artificial electron donors. For instance, cells of *Th. roseopersicina* in the presence of methyl viologen and dithionite evolve H_2 corresponding to 400 ml h^{-1} g^{-1} dry biomass [21]. This process is catalyzed by hydrogenase and is independent of illumination.

3.2 Green Bacteria

Green bacteria, similar to the purple ones, perform anoxygenic photosynthesis [165, 216, 223]. Most species are strict anaerobes and obligate photoautotrophs. They use inorganic sulfur compounds and molecular hydrogen as electron donors [163, 164, 169, 176]. Only some green bacteria are able to grow both in the light and in the dark under aerobic heterotrophic conditions [164, 169, 226, 227].

Some species reveal a capability for N_2 fixation [45, 165, 174].

Evidence of H_2 evolution by green bacteria is very scanty and concerns the members of two genera: *Chlorobium* and *Pelodictyon* [45, 228]. According to available data H_2 production by these microorganisms occurs in the light and is poor whereas H_2 production by some mixed cultures containing green bacteria is intensive [21, 228]. Thus, the capacity of green bacteria for H_2 formation requires further investigations.

3.3 Cyanobacteria

Cyanobacteria (= blue green algae) represent the most numerous group of phototrophic prokaryotes [24, 71, 230-232]. In contrast to purple and green bacteria, they possess two photosystems, use H_2O as electron donor and evolve oxygen. But some cyanobacteria may grow in the presence of organic compounds without the action of the second photosystem [24, 231]. The ability of some cyanobacteria for anoxygenic photosynthesis under anaerobic conditions in the presence of sulfide as electron donor has also been reported [232-234].

Several cyanobacteria consume molecular hydrogen. But according to available data the oxidation of H_2 inhibits their growth [54, 55, 236-240]. Some cyanobacteria are able to grow under heterotrophic conditions in the dark [71, 231, 232]. A considerable number of cyanobacteria may fix N_2 [231, 232, 240-242].

Certain cyanobacteria produce H_2 in the dark. But under such conditions H_2 evolution is usually poor [50, 243-248].

In the light, cyanobacteria produce H_2 more intensively. Such a capacity was discovered for a rather great number of species [8, 12, 21, 54, 55, 243, 244, 249-272] belonging to different genera (Table 3). But not all cyanobacteria cultures evolving H_2 were pure [262].

Photoevolution of H_2 also occurs in *Anabaena azollae* growing as symbiont of water fern *Azolla* and in pure culture [262-264].

Most of the cyanobacteria capable to produce H_2 in the light are filamentous. But there is also evidence of their capability to photoevolve H_2 by unicellular cyanobacteria [262].

All known cyanobacteria, except *Oscillatoria limnetica* [255, 256] which produce H_2 in the light, may use N_2 as nitrogen source [8, 21, 54, 55]. The most detailed investigations of H_2 metabolism concern *Anabaena cylindrica* [8,21,54,243,250,254,258,259,266]. The results of these studies have allowed to discover the capacity of cyanobacteria for H_2 photoevolution [243,249]. Among cyanobacteria producing H_2 there are fresh water and marine microorganisms [21,246].

Usually, H_2 production is studied in cell suspensions [8,21]. But there are some data about H_2 evolution simultaneously with protein synthesis [261]. On the other hand, it is known that H_2 may be produced by immobilized cells of some cyanobacteria [8,270,271].

Similar to purple bacteria the rate of H_2 evolution in cyanobacteria depends on the intensity of light. For instance, *An. cylindrica* strain B269 produces more H_2 at 4000 lx than at 700 lx whereas at low light intensity the production of H_2 may continue for a longer time [8,21,223,270-272].

Solar irradiation results in higher rates of H_2 formation than the use of artificial light. But the efficiency of energy conversion is lower. In the first case, it corresponds to 0.1–0.2%, in the second one it reaches 2.7% [8,223,236,254,268,273].

Alongside with mesophilic cyanobacteria there are known thermophilic species capable to produce H_2 at 45–54 °C [258,259]. The content of iron in the growth media is important for H_2 production by cyanobacteria [8,272] as well as by other organisms.

In contrast to purple bacteria, photoevolution of H_2 by several cyanobacteria is possible in the presence of oxygen. But in many cases these microorganisms produce more H_2 in an argon atmosphere or in the presence of 1–10% O_2. In cyanobacteria not containing heterocysts the process of H_2 formation is more sensitive to oxygen [8,54,223,259,270,271,274,275]. *O. limnetica* adapted to anoxygenic photosynthesis is able to evolve H_2 but only under anaerobic conditions [239,254,255].

The presence of considerable amounts of ammonium or nitrates in the medium represses H_2 photoevolution by cyanobacteria capable to assimilate N_2. But at very low concentration ($\sim 10^{-4}$ M) NH_4^+ may somewhat prolong H_2 production by these microorganisms and increase its rate [8,21,222,274,275].

Molecular nitrogen inhibits H_2 photoevolution by N_2-fixing of cyanobacteria [8,21,50,222,243,249,253,258,259]. In some cases, the presence of 15% N_2 is sufficient to complete inhibition of this process. However, at a low concentration, N_2 as well as NH_4^+ stimulate H_2 production by some cyanobacteria [8,222,270].

The inhibitory effect of molecular nitrogen on photoevolution of H_2 by *An. cylindrica* and other cyanobacteria is relieved to a considerable degree by CO. Moreover, in the presence of CO and acetylene (at a definite ratio) the rate of H_2 photoevolution by cyanobacteria both in the air and in the Ar atmosphere may be even increased [8,253,266,267,269,275-277].

Sometimes, H_2 photoevolution by cyanobacteria, similar to some N_2-fixing chemotrophs [82,83], was revealed only in the presence of CO [262] or carbon monoxide and acetylene (see Sect. 3.4) [8,51,266,267]. But acetylene without CO inhibits photoevolution of H_2 [54,55,274,278].

All these data show that H_2 photoproduction by cyanobacteria capable to fix N_2 proceeds with participation of nitrogenase [8,24,54,222,243,254]. In some cases, nitrogenase catalyzes H_2 production by cyanobacteria also in the dark [243,244,247].

But photoevolution of H_2 in *O. limnetica*, not capable of N_2 fixation, is catalyzed by hydrogenase [255,257]. In other cyanobacteria hydrogenases [21,50,51,237,239,254,279-283] seem to participate principally in H_2 oxidation but not in its photoevolution. Only in the presence of reduced methyl viologen, glucose, pyruvate, or other organic compounds do some cyanobacteria produce H_2 with participation of hydrogenases. This process is independent of illumination [8,12,21,54,55,243].

Cyanobacteria may produce H_2 in the light on media not containing any organic or inorganic electron donors except water. This fact shows photoevolution of H_2 by cyanobacteria to be coupled with the use of H_2O as electron donor. The participation of H_2O in the photoproduction of hydrogen in cyanobacteria is confirmed by simultaneous formation of H_2 and O_2 in the light [8,54,274]. But several experimental data on *An. cylindrica* and some other cyanobacteria reveal that water is not a direct electron donor for H_2 evolution by these microorganisms. In *An. cylindrica* and many other cyanobacteria H_2 photoevolution proceeds in special types of cells (heterocysts) where nitrogenase is localized. Heterocysts are defective in the photosystem II. They are unable to use H_2O as electron donor and do not evolve O_2. But they are susceptible to photophosphorylation and active respiration [231-233].

Probably, organic compounds function as electron donors for H_2 production in heterocysts. Such compounds may be formed in vegetative cells from CO_2 and H_2O and transported into heterocysts [54,229,231,233,259,274]. But the pathway of electron flow from organic compounds to nitrogenase is not yet defined.

Thus, it is supposed that water is involved in H_2 photoevolution in *An. cylindrica* and other N_2-fixing cyanobacteria. But CO_2 participates in this process as well:

$$H_2O + CO_2 \xrightarrow{\text{light}} \begin{matrix}\text{Organic}\\ \text{compounds}\end{matrix} \xrightarrow{\text{light}} H_2$$

This assumption does not exclude, however, that in some cyanobacteria H_2O may serve as direct electron donor for H_2 photoproduction:

$$H_2O \xrightarrow{\text{light}} H_2 + 0.5\ O_2$$

Of particular interest in this respect are non-heterocystions cyanobacteria capable of nitrogen fixation in the presence of O_2 [24,231,232].

It should also be noted that among the phototrophs only cyanobacteria are capable to form H_2 under aerobic conditions. Therefore, studies on the H_2 production of these microorganisms as a result of biophotolysis of water as well as on the photoassimilation of N_2 attract great attention [3-12,21,52,54,60-63,222].

In a rather short time, considerable work has been done for selecting cyanobacterial strains capable of extensive H_2 photoproduction. Besides *An. cylindrica* B629 [8,12,21,44,54,274], most active strains are represented by thermophilic cyanobacteria [258,259] and some other species (Table 4). The rate of H_2 photoproduction by *An. cylindrica* reaches 30–40 ml h^{-1} g^{-1} dry biomass. A similar rate of H_2 evolution is observed for *Mastigocladus laminosus* and an even higher in *Mas. thermophilus* [257-259]. *An. cylindrica* and some other cyanobacteria may photoproduce H_2 during 30 and even more days [8,12,54,222,254,258,285].

Thus, the results of studies on the H_2 production of cyanobacteria are rather favourable.

H_2 formation is also possible in two-stage systems containing cyanobacteria and purple bacteria cells [8,222]. A method of converting solar energy into electricity using cyanobacteria cells is being developed [287].

3.4 Algae

The capacity for H_2 evolution has been discovered in several algae belonging to different systematic groups [8,21,31,33,47–52,260,288–292]. Especially numerous are those of green algae (Table 3).

The majority of algae capable to produce H_2 belong to the microforms, but some macro-algae are also known [291]. Representatives are red algae and the green alga *Codium fragilis*.

Similar to higher plants and cyanobacteria algae perform oxygenic photosynthesis [293]. But some of these organisms have the capacity for CO_2 photoreduction using H_2 as electron donor [46–48,290]. Such algae are also able to oxidize molecular hydrogen in the dark in the presence of O_2 or other electron acceptors [46–48].

However, no growth of these organisms is observed under these conditions. Only in the presence of some organic compounds and O_2 may several algae grow in the dark [46–48,293].

The formation of H_2 by algae only occurs in cell suspensions. This process generally requires a preincubation of algae under anaerobic conditions [46,289,294–297]. The phase of culture growth is also of some importance [296].

Most algae are able to evolve H_2 both in the dark and in the light. But in the presence of light the rate of this process is usually higher [8,46–48,289,290,298–300]. According to available data only in red algae does H_2 production not depend on the presence of light [291].

The intensity of light required for maximum H_2 production is much lower than that for CO_2 assimilation [298]. For instance, in *Chlorella pyrenoidosa* photoproduction of H_2 reaches its maximum rate at 0.75 mW cm^{-2} (7.5×10^3 erg cm^{-2} s^{-1}) [299]. In *Ch. vulgaris* the optimum light intensity for this process is lower and depends on the composition of the medium [300]. But mutants of some algae containing only small quantities of chlorophyll and being uncapable of photosynthesis may produce H_2 at a very high light intensity [301]. The action spectrum of H_2 formation in algae is similar to that of O_2 evolution. But H_2 production is possible as well when cells are irradiated with a red light which is not effective for photoassimilation of CO_2 and evolution of O_2 [50,303,304–307].

It was also shown that *Scenedesmus obliquus* grown under heterotrophic conditions produced H_2 at a higher rate than when grown in a mineral medium [290].

In contrast to purple bacteria and many cyanobacteria, H_2 production in algae does not depend on the source of nitrogen and is not inhibited by N_2 or NH_4^+ [8,21,46,222,289]. But in the presence of considerable amounts of H_2 as well as in the presence of CO_2 and CO this process is inhibited [8,21].

It is known for a long time that H_2 production by some algae in the dark increases in the presence of organic substrates in the media. In the light H_2 formation by algae is possible in the absence of exogenic organic compounds. But in their presence the rate of H_2 evolution is often increased too. In general, such effect on H_2 evolution by algae in the dark and in the light is caused by glucose. But in certain algae H_2 formation is increased in the presence of acetate, pyruvate, lactate, and other organic compounds [46, 289, 295, 298, 300, 307].

The rate of H_2 formation by algae is not high (Table 4). In the dark it corresponds to 0.2–2.0 and in the light to 1.5–5.5 ml h^{-1} g^{-1} dry biomass. Only in a few algae is the rate of H_2 production higher. For instance, a thermophilic strain of *Ch. vulgaris* produces up to 43 ml H_2 h^{-1} g^{-1} dry biomass in the presence of organic substrates [295].

The maximum rate of H_2 formation by some algae corresponds to 48–70 ml h^{-1} g^{-1} chlorophyll [222, 292, 308]. This value is lower than the rate of photoassimilation of CO_2 and of O_2 evolution [8]. However, the efficiency of the conversion of light energy in the H_2 production may reach 10% [8, 32]. The period during which H_2 is formed most rapidly by algae is only a few hours. After about one day or less this process usually ceases [290, 298, 299, 303, 308]. The main cause of a rapid inhibition of H_2 photoproduction by algae is the sensitivity of this process to O_2. Therefore, H_2 production by algae is observed at low intensity of light when O_2 evolution proceeds very slowly [299, 301–303].

In the dark H_2 formation by algae similarly to other organisms is a result of degradation of endogenous or exogenous organic compounds. It is coupled with the synthesis of some other products: formate, acetate, ethanol, glycerol, and CO_2 [46–48, 311, 312]. But metabolic pathways of organic substrates leading to H_2 evolution need further investigations.

The mechanism of photoevolution of H_2 in algae seems to be still more complicated. Several data show H_2 photoproduction in algae to be also linked to the oxidation of endogenous or exogenous organic compounds [46, 50, 298–300, 310].

Two possible metabolic pathways of organic substrates coupled with H_2 photoproduction are discussed. According to the first supposition organic compounds play the role of electron donors on the level of photosystem I or photosystem II [46, 47, 312, 313]. Such a hypothesis is based on the absence of the sensitivity of H_2 photoevolution in algae to the action of uncouplers [222, 298, 307, 308].

According to the second hypothesis, H_2 photoproduction in algae occurs as a result of reversed electron transport from organic substrates. The necessary energy for this process may be provided by the function of photosystem I [50, 299, 300, 303–305].

But there are some data according to which photoproduction of H_2 in algae may result from the action of two photosystems and participation of water as electron donor [8, 21, 46]. Recent experimental evidence confirms the possibility of H_2 photoevolution in algae as a result of biophotolysis of water [8, 222, 243, 290, 314, 316, 317].

After incubation in the dark *Sc. obliquus* and some other algae may photoproduce H_2 and evolve O_2 in the ratio of 1.94. Mutants of *Sc. obliquus* and *Chlamydomonas reinhardii* with defects in the second photosystem do not produce H_2 in the light or the rate of hydrogen evolution is very low [8, 222, 290, 308, 312, 316, 317].

Ch. vulgaris and *Sc. obliquus* may produce H_2 in the light for a rather long time if O_2 is scavenged. For oxygen removal dithionite and some other compounds may be used. But in the presence of dithionite and DCMU H_2 photoevolution ceases [222, 316, 317]. In some other algae DCMU also inhibits H_2 photoproduction at the beginning of their irradiation [243, 290, 291].

Thus, all data show that H_2 production in algae may occur due to the oxidation of some organic compounds and biophotolysis of H_2O. In all cases, H_2 production is catalyzed by hydrogenases [8,21,46,48]. But H_2 production by algae is insufficient for practical applications. Further work for the selection of suitable strains and for the stabilization of the process is needed.

4 Enzymes Catalyzing H_2 Production

4.1 Hydrogenases

Hydrogenases are present in the majority of the microorganism producing molecular hydrogen and in all species oxidizing it [8,10,18,21]. All known hydrogenases (EC 1.12) are Fe-S-containing proteins catalyzing the reversible reaction

$$H_2 \rightleftharpoons 2\,H^+ + 2e^-$$

Hydrogenases of various microorganisms differ in molecular weight, the electron donor (acceptor) they interact with, cellular localization and in other properties. Some

Table 5. Localization of hydrogenases in different microorganisms

Microorganisms	Soluble	Membrane-bound	Ref.
Clostridium pasteurianum[a]	+	—	108)
Desulfovibrio gigas[b]	+	—	329)
D. vulgaris M	—	+	361)
D. vulgaris[b]	+	—	330)
D. desulfuricans	—	+	362)
D. desulfuricans	+	+	331)
Desulfotomaculum nigrificans	—	+	366)
Dm. ruminis	—	+	367)
Megasphaera elsdenii	+	—	369)
Methanobacterium thermoautotrophicum	+	—	370)
Veillonella alcalescens	+	—	365)
Azotobacter vinelandii	—	+	371)
Alcaligenes eutrophus	+	+	363)
A. latus	—	+	152)
A. paradoxus	—	+	363)
A. ruhlandii	+	+	368)
Aquaspirillum autotrophicum	—	+	152)
Arthrobacter sp.	—	+	152)
Azospirillum lipoferum	—	+	541)
Xanthobacter autotrophicus	—	+	152)
X. flavus	—	+	541)
Derxia gummosa	—	+	541)
Escherichia coli	—	+	324, 364)
Flavobacterium autothermophilum	—	+	152)
Mycobacterium gordone	—	+	152)
Mycrocyclus aquaticus	—	+	541)
Nocardia opacal[b]	+	—	36)
N. autotrophica	+	—	152)

Table 5 (continued)

Microorganisms	Soluble	Membrane-bound	Ref.
N. autotrophica	+	—	152)
Proteus mirabilis	—	+	373)
Paracoccus denitrificans	—	+	373)
Renobacter vacuolatum	—	+	541)
Pseudomonas methylica	+	+	374)
Ps. thermophila	—	+	375)
Ps. flava	—	+	152)
Ps. saccharophila	+	+	152)
Ps. facilis	—	+	376)
Ps. hydrogenovora	—	+	152)
Ps. palleronii	—	+	363)
Ps. carboxydovorans	—	+	152)
Ps. pseudoflava	—	+	384)
Rhizobium japonicum	—	+	541)
Rhodospirillum rubrum	+	+	182)
R. rubrum S1	—	+	377)
R. rubrum G9[c]	+	—	333)
Rhodopseudomonas capsulata	—	+	383)
Chromatium sp.	—	+	382)
Chr. vinosum	—	+	381)
Thiocapsa roseopersicina BBS	+	+	300)
Anabaena cylindrica[d]	+	+	21)
Nostoc muscorum[d]	+	+	379)
Spirulina maxima	+	—	349)

Note. Hydrogenase was found: a — both in the periplasm and in the cytoplasm; b — in the periplasm; c — in the culture medium; d — in the fractions of the vegetative cells and in the heterocysts

microorganisms contain two hydrogenases differing in a number of properties and functions [10, 17, 21, 36, 108, 323, 525, 526].

Hydrogenase activity of whole cells and cell-free preparations may be evaluated from the rate of reduction of various H_2 acceptors [17, 21, 318], H_2 evolution in the reaction with different electron donors [31], catalysis of exchange reactions [17], the reaction of ortho-para hydrogen conversion [315] and electrochemically [320–322]. Hydrogenase activity is most precisely assayed by the H_2 uptake method in the presence of certain electron acceptors and by exchange reactions [17, 18].

Methods of purification of hydrogenases are described in a number of reviews [10, 17, 18, 21, 36].

4.1.1 Localization and Functions

All hydrogenases are commonly subdivided into soluble and membrane-bound types (Table 5).

In some cases hydrogenase activity is observed in broken cells in different fractions [328]. Some data confirm that hydrogenases differing in certain properties and functions may be localized in the same cellular fraction [279, 280].

In *C. pasteurianum* [107, 108] and some other bacteria [329–332] a considerable amount of hydrogenases was located within the periplasmic space. It was also observed that *R. rubrum G-9* [333], *Desulfovibrio vulgaris* [334] and *D. desulfuricans* [335] partly released hydrogenases into the culture medium.

In some microorganisms various processes may be carried out by one hydrogenase. However, in the species able to synthesize two types of hydrogenases they very probably fulfil different functions [10, 18, 21]. In *A. eutrophus* the soluble hydrogenase ensures utilization of H_2 for NAD reduction which is essential for CO_2 assimilation via the ribulosebisphosphate cycle. In contrast, the membrane-bound hydrogenase of *A. eutrophus* catalyzes the electron transport from H_2 to the respiratory chain [36]. A number of other H_2-oxidizing chemotrophic bacteria and phototrophic bacteria contain two types of hydrogenases (Table 5).

From the available data it may be concluded that membrane-bound hydrogenases more frequently catalyze the consumption of H_2 by microorganisms. The soluble hydrogenases are often involved in H_2 evolution. However, in several cases (e.g., *E. coli*) the evolution of H_2 is due to a membrane-bound hydrogenase. On the other hand, the soluble hydrogenases (e.g., *A. eutrophus*) may catalyze H_2 oxidation [36, 364].

4.1.2 Composition and Structure

Homogeneous preparations of hydrogenase have been obtained from several microorganisms capable of both production and consumption of H_2 (Table 6).

In comparison with soluble hydrogenases ($M = 45–89 \times 10^3$), the majority of the enzymes isolated from the membrane fractions possess higher molecular weights ($62–203 \times 10^3$). An exception is the soluble hydrogenase from *A. eutrophus* with a molecular weight of 205000 [319, 390]. The molecular weight of membrane-bound hydrogenase of this bacterium is 98000 [152, 391].

A considerable number of purified hydrogenases turned out to be monomeric (Table 6). However, the soluble hydrogenases from *D. gigas* [329, 387], *Chr. vinosum* [393], *Th. roseoperisicina* [380], and the membrane-bound hydrogenases from *D. vulgaris* [361, 386] and other bacteria [152, 382, 384, 428] consist of two subunits. The most complex subunit composition possess the soluble hydrogenase from *A. eutrophus* [319], the membrane-bound hydrogenases from *Proteus mirabilis* [372] and *Par. denitrificans* [373]. The enzymes consist of four subunits (Table 6).

All studied hydrogenases from various microorganisms [329, 331, 380, 382, 385, 429, 430] are characterized by the predominant occurrence of acidic amino acids. This is in correspondence with the isoelectric points of the enzymes which lie in the pH range from 4.15 to 6.02. Characteristic of many hydrogenases is a high content of aromatic and hydrophobic residues [18, 21].

The catalytic activity is mainly determined by the iron-sulfur cluster which is included (according to EPR and optical spectra) into the composition of all hydrogenases [329, 378, 433, 434]. Such a cluster is required for the manifestation of hydrogenase activity [10, 18, 31, 36, 108, 432]. The four-fold multiplicity of iron and sulfur atoms, the equality between the number of S^{2-}-groups and iron atoms, and the data concerning the transfer of iron and sulfur atoms from hydrogenases to artificial Fe_4S_4 compounds in the exchange reaction [10, 435] support such a conclusion.

Table 6. Properties of homogeneous hydrogenase preparations from various microorganisms

Nr.	Properties	*Clostridium pasteurianum* W5 [108, 378]	*Desulfovibrio vulgaris* Hildenborough [330]	*D. vulgaris* NSIB [385]	*D. vulgaris* Miyazaki [361, 386]	*D. gigas* [324, 387]
1.	Localization	Cytoplasm	Periplasm	Cytoplasm	Membranes	Periplasm
2.	ε (mM^{-1} cm^{-1})	$\varepsilon_{280} = 60$ $\varepsilon_{400} = 25$	n.d. $\varepsilon_{400} = 46$	$\varepsilon_{277} = 41$ $\varepsilon_{408} = 3.1$	$\varepsilon_{280} = 164$ $\varepsilon_{400} = 47$	$\varepsilon_{280} = 170$ $\varepsilon_{400} = 47$
3.	M_r	60,500	49,000	45,400	89,000	89,500
4.	Subunits	1 × 60,500	1 × 49,000	n.d.	1 × 28,000 1 × 59,000	1 × 62,000 1 × 26,000
5.	Fe^{2+} per mol	12	12	0.7	8	12
6.	Labile S^{2-} per mol	12	12	0.35	8	12
7.	Isoelectric point	5.0	n.d.*)	7.0	6.2	7.0
8.	EPR signals: oxidized form	Hipip type, g > 2	n.d.	n.d.	n.d.	Hipip type
9.	Reduced form	g = 2.099; 1.961; 1.892	g = 2.3; 1.86	n.d.	n.d.	EPR-silent
10.	Type of Fe-S centre	4Fe-4S	n.d.	n.d.	2 × [4Fe-4S]	3 × [4Fe-4S]
11.	Natural electron donor/acceptor	Ferredoxin, flavodoxin	Cytochrome c_3	Cytochrome c_3	Cytochrome c_3	Cytochrome c_3/flavodoxin
12.	Physiological function	H_2 evolution	H_2 oxidation	H_2 oxidation	H_2 oxidation	H_2 oxidation

Table 6 (continued)

Nr.	*D. desulfuricans* [331]	*Megasphaera elsdenii* [369, 388]	*Escherichia coli* [10]	*Proteus mirabilis* S503 [372]	*Paracoccus denitrificans* DSM381 [389]	*Pseudomonas pseudoflava* DSM1034 [384]
1.	Membranes	Cytoplasm	Membranes	Membranes	Membranes	Membranes
2.	n.d.	$\varepsilon_{275} = 143$	$\varepsilon_{280} = 140$	$\varepsilon_{280} = 390$	n.d.	n.d.
	n.d.	n.d.	$\varepsilon_{400} = 49$	$\varepsilon_{400} = 106$	n.d.	n.d.
3.	52,000	50,000	113,000	205,000	63,000	98,000
4.	1 × 52,000	1 × 50,000	2 × 56,000	2 × 63,000 2 × 33,000	1 × 63,000	1 × 65,000 1 × 30,000
5.	12	12	12	24	n.d.	6
6.	n.d.	12	12	24	n.d.	6
7.	6.0	7.0	4.2	4.75	n.d.	6.5
8.	n.d.	Hipip type	Hipip type, $g > 2$, axial	Hipip type, $g > 2$, axial	n.d.	n.d.
9.	n.d.	8Fe Ferredoxin type	EPR-silent	EPR-silent	n.d.	n.d.
10.	4Fe-4S	n.d.	n.d.	n.d.	n.d.	n.d.
11.	n.d.	Ferredoxin	n.d.	n.d.	n.d.	n.d.
12.	H_2 oxidation	H_2 evolution	H_2 evolution	H_2 evolution	n.d.	n.d.

Table 6 (continued)

Nr.	*Alcaligenes eutrophus* H16 [319, 390]	*A. eutrophus* H16 [152, 391]	*Rhodospirillum rubrum* S1 [392]	*Chromatium* sp. D [382]	*Chromatium vinosum* [381]	*Thiocapsa roseopersicina* BBS [21, 214, 380, 396]
1.	Cytoplasm	Membranes	Membranes	Membranes	Cytoplasm	Cytoplasm and membranes
2.	$\varepsilon_{280} = 250$ $\varepsilon_{420} = 60$	$\varepsilon_{280} = 122$ $\varepsilon_{408} = 30$	$\varepsilon_{280} = 36$ $\varepsilon_{400} = 8.3$	$\varepsilon_{280} = 96$ $\varepsilon_{410} = 14$	n.d. n.d.	n.d. n.d.
3.	20,5000	98,000	66,000	98,000	68,000	68,000
4.	1 × 68,000 2 × 29,000 1 × 60,000	1 × 67,000 1 × 31,000	1 × 66,000	2 × 50,000	2 × 35,000 4 × 20,000(?)	1 × 47,000 1 × 25,000
5.	12	6	4	4	n.d.	4
6.	12	6	4	4	n.d.	4
7.	4.85	6.5	5.3	4.2 and 4.4	7.0	4.15 and 4.2
8.	Weak signal, g = 2.02	n.d. n.d.	Hipip type, g > 2, axial	Hipip type, g > 2, axial	n.d.	Hipip type, g = 1.98; 2.03
9.	High temp., g = 2.04; 1.95 low temp., g = 2.04; 2.00; 1.95; 1.93; 1.86	n.d.	EPR-silent	EPR-silent	g = 1.94; 1.96; 1.98 2.03	EPR-silent
10.	2 × [4Fe-4S] 2 × [2Fe-2S]	n.d.	n.d.	n.d.	n.d.	4Fe-4S
11.	NAD	Quinone(?)	n.d.	n.d.	n.d.	Cytochromes (c_3, c_3' and c_{552})
12.	H_2 oxidation	H_2 oxidation	H_2 oxidation	H_2 oxidation	H_2 oxidation	H_2 oxidation

*) n.d. = not determined (in all tables).

4.1.3 Interactions of Electron Donors and Acceptors with Hydrogenases

Depending on the nature of the electron carrier, hydrogenases interact in vivo. These enzymes are classified as H_2-ferredoxin oxidoreductase (EC 1.12.7.1), characteristic of *Clostridia* [108] and *M. elsdenii* [388], H_2-ferricytochrome oxidoreductase (EC 1.12.2.1), found in sulfate-reducing bacteria [430], and H_2: NAD oxidoreductase (EC 1.12.1.2) characteristic of *A. eutrophus* [319].

The nature of the electron donors (acceptors) which in teract with hydrogenases may be different (Table 6). Thus, for *Methanobacterium thermoautotrophicum* hydrogenase, the role of the physiological electron carrier is probably played by the factor F_{420} [370]. According to our results, c-type cytochromes are the physiological electron donors (acceptors) in the purple bacteria (*Rh. capsulata, Ec. shaposhnikovii, Th. roseopersicina*) and in the cyanobacterium *A. cylindrica* [21,327,332]. The mentioned cytochromes were isolated from the corresponding bacteria. However, for a number of

Table 7. Activity of homogeneous hydrogenase preparations from various microorganisms

Enzyme sources	Purification (-fold)	Electron donor (acceptor)	Activity, μmol min^{-1} mg^{-1} protein		Ref.
			H_2 evolution	H_2 oxidation	
			Soluble hydrogenases		
Clostridium pasteurianum	320	MV	317.0	317.0–1660.0	31, 394)
		MV + Fd	3960.0	n.d.	
		MB	0	960.0	
C. pasteurianum	200	MV	268.3	n.d.	21)
		Fd *C. pasteurianum*	409.3	n.d.	
Desulfovibrio vulgaris	15	MV	n.d.	64.0	385)
		BV	n.d.	53.0	395)
	360–440	MV	3800.0	3800.0	330)
D. desulfuricans	n.d.	MV	3800.0–9000.0	n.d.	331)
		BV	1050.0	n.d.	
Alcaligenes eutrophus H16	68	MV	48.0	n.d.	319)
		NADH	1.2	n.d.	
Thiocapsa roseopersicina	52	MV	60.0–612.0	21.2–260.0	380, 396)
		Fd	0.2	n.d.	
		cytochrome c_3'	1.8	n.d.	
		D_2-H_2O	n.d.	0.4	
		Membrane-bound hydrogenases			
Desulfovibrio vulgaris	n.d.	MV	200.0	610.0	361, 397)
Pseudomonas pseudoflava	58	MB	n.d.	171.9	384)
Paracoccus denitrificans	109	BV	n.d.	5.1	389)
Rhodospirillum rubrum	490	MV	25.9	n.d.	392)
Chromatium sp.	1700	MV	35.0	n.d.	382)
		MB	n.d.	330.0	
		H_2-D_2O	n.d.	82.1	
Chr. vinosum	280	MV	1.9	n.d.	381)
Thiocapsa roseopersicina	67	MV	3.4–50.0	44.0	214, 380)
		D_2-H_2O	n.d.	0.4	

microorganisms the nature of the physiological electron donors (acceptors) interacting with hydrogenases is not yet established [18,21,36,541].

The majority of hydrogenases catalyze in vitro the reversible oxidation-reduction reaction of some artificial and natural electron carriers (Table 7).

Exception are the so-called "unidirectional" hydrogenases from *C. pasteurianum* [108] and some other bacteria [21,84,370,405,436,528] which mainly catalyze H_2 consumption. The rate of H_2 evolution, catalyzed by highly purified preparations of such hydrogenases [21,405,436] does not exceed 1–10% of the hydrogen-oxidizing activity of the enzyme.

Generally, the rates of the reactions catalyzed by hydrogenase are much higher when they correspond to that in the direction of the electron flow in vivo.

The activity of the hydrogenases of several microorganisms [21,330,387,430] increases considerably when two electron donors are present in the reaction mixture. For instance, H_2 evolution by *C. pasteurianum* hydrogenase is most pronounced in the presence of MV and ferredoxin from spinach (Table 7). The highest rate of H_2 evolution (3800–9000 μmol min^{-1} mg^{-1} protein) in the presence of reduced artificial (MV) or natural (ferredoxin) electron donors is obtained with the soluble hydrogenases from *C. pasteurianum* and *D. desulfuricans* (Table 7). Oxidation of hydrogen proceeds at a higher rate with the membrane-bound hydrogenases.

4.1.4 Mechanism of Catalysis

As a rule, the initial stationary rates of H_2 evolution and consumption involving hydrogenase in the presence of artificial electron carriers fit nicely the Michaelis-Menten equation. The Michaelis constants obtained with different electron acceptors (donors) lie in the range of 10^{-6}–10^{-2} M [18,21]. The lowest K_m value (1.4×10^{-6} M) determined for H_2 consumption is characteristic of *Rhizobium japonicum* hydrogenase [405] and the highest value (1.5×10^{-2} M) of the *Nocardia opaca* and *M. elsdenii* hydrogenases. For the hydrogenases of *C. pasteurianum* [394] and *Th. roseopersicina* [440] it is shown that K_m decreases with increasing pH.

The K_m values of the redox reactions of natural electron carriers, catalyzed by hydrogenases, are lower than those of redox reactions with artificial carriers [21]. For hydrogenases of clostridia the reaction rate is considerably higher with the so-called "bacterial-type" ferredoxins (two-electron carriers) than with plant ferredoxins (one-electron carriers).

On the basis of the results of kinetic studies of the hydrogenase-catalyzed reactions, the cluster model of the active center of the enzyme and other data it may be assumed that the action of the hydrogenase involves a heterolytic cleavage of the hydrogen molecule, proceeding in two stages [10,394,437,441,442].

1) Activation of molecular hydrogen and transfer of two electrons from the hydride to the active center of the enzyme [10,382,437]:

$$E + H_2 \rightleftharpoons E{:}H^- + H^+$$

2) Electron transfer from the enzyme to an acceptor A^+. Depending on the nature of the electron acceptor, this step apparently proceeds either via a hydride ion (H^-) transfer [439]

$$E{:}H^- + B^+ \rightleftharpoons E + BH\ ,$$

or via an electron transfer [440]:

$$E{:}H^- + 2A^+ \rightleftharpoons E + 2A + H^+$$

Thus, hydrogenases may catalyze both the reduction of electron acceptors by hydrogen and H_2 evolution with various donors.

4.1.5 Stability

4.1.5.1 Stability to Oxygen

A significant diversity in the degree of the sensitivity to oxygen is observed among hydrogenases from different microorganisms [10,18,21,31]. Many hydrogenases are very sensitive to O_2. A low stability to oxygen is typical of the hydrogenases from clostridia and algae. The half-life of hydrogenase inactivation during aerobic storage of the crude cell extracts of different clostridia is 15–20 h [336]. The highly purified *C. pasteurianum* hydrogenase was inactivated completely and irreversibly after 5–20 min storage in the air [108,443]).

More stable to O_2 inactivation are the hydrogenase preparations of several sulfate-reducing [331] and purple bacteria [18,21,392]. The hydrogenase of *D. desulfuricans* retains 100% activity on storage in air (4 °C, 160 mM phosphate buffer, pH 7.5) for a month.

The half-life of *Th. roseopersicina* hydrogenase stored in air in the form of partially purified preparations at 4 °C is about 6 d [444,445]. Much more stable to O_2 are the highly purified preparations of this hydrogenase. The half-life periods of such preparations in air at 4 and at 24 °C are about 190 and 60 d, respectively. A higher hydrogenase rate of inactivation in the case of cell-free extracts of *Th. roseopersicina* (also that of *E. coli* [446] and *A. eutrophus* [447]) is likely to be related to the action of superoxide radicals or other products of oxygen reduction which may be generated by some cellular components present in the preparation [448].

According to our data, *T. roseopersicina* hydrogenase reveals a high stability to H_2O_2 but is inactivated much faster in the presence of the $O_2^{\dot{-}}$-regenerating system, and especially in the presence of 5 mM H_2O_2 and of the $O_2^{\dot{-}}$-regenerating system.

In the presence of Fe- and Mn-POD and, to a lesser extent, of albumin the *Th. roseopersicina* hydrogenase inhibition by peroxide anions and H_2O_2 decreased to 25%. But *Rh. capsulata* hydrogenase is inactivated much more rapidly in the presence of H_2O_2 than in the presence of $O_2^{\dot{-}}$. However, similar to *Th. roseopersicina* hydrogenase, the inhibitory influence of H_2O_2 and $O_2^{\dot{-}}$ on the hydrogenase activity of *R. capsulata* enzyme is completely removed in the presence of Fe-SOD and catalase.

The oxygen-inactivated *Th. roseopersicina* hydrogenase like that of the sulfate-reducing bacteria [400,449] recovers in the presence of reducing agents ($Na_2S_2O_4$) and under anaerobic conditions [380]. However, in contrast to the hydrogenases of purple bacteria, the stability of *D. vulgaris* hydrogenase towards oxygen is affected neither by catalase nor by peroxide dismutase [334].

The hydrogenases from cyanobacteria *An. cylindrica* and *Spirulina platensis* are less stable to oxygen [18], the half inactivation period is about 24 h on storage in air at +4 °C.

The inactivating action of O_2 is extremely pronounced for the hydrogenases of green algae. In these organisms under aerobic conditions hydrogenases are preserved in an inactive oxidized state; they turn into an active reduced state during anaerobic incubation [18,46,294,450]. The hydrogenase of *Chlamydomonas reinhardii* is inactivated irreversibly during 2–3 min storage of cell-free extract in air at 4 or 20 °C. However, data exist which reveal that the hydrogenase of *Chl. reinhardii* mutant is more stable to O_2 [451].

Among the hydrogenases studied the enzymes from certain hydrogen bacteria are most stable to O_2 [18,402,452]. Moreover, storage of the soluble hydrogenase from *A. eutrophus* in the presence of reducing agents ($Na_2S_2O_4$, NADH, mercaptoethanol) reduces its stability; high concentrations of these substances lead to a rapid inactivation of the enzyme [363,453–455]. The rate of inactivation of hydrogenase preparations from *A. eutrophus* is 3–4times lower in air than under anaerobic conditions.

This is probably related to the existence of two states of the enzyme, differing in their activity and stability to various factors. These states may correspondingly be the hydrogenases with oxidized and reduced active centers. The hydrogenase in the oxidized state is not catalytically active, but it is more stable. It may be activated by the addition of reducing agents whereas in a reduced state it is more unstable.

Furthermore, the decrease of the *A. eutrophus* hydrogenase activity in air in the presence of H_2 or NADH may be related to the inactivation of the enzyme by O_2^- which is generated by this hydrogenase [447,541]. The same may be true in the case of *Rh. capsulata* and *Th. roseopersicina* hydrogenases. This probably explains the decrease of H_2 evolution in air catalyzed by *A. eutrophus* Z1 hydrogenase when NADH is used as an electron donor [455].

4.1.5.2 Heat Stability

The activation energies of the exchange reactions catalyzed by hydrogenases differ from those of the reactions of H_2 consumption or evolution. In the first case, the mean value of the activation energy is about 7.5 ± 0.1 kcal mol^{-1}, and in the second- 15 ± 1 kcal mol^{-1} [18,21]. The majority of hydrogenase-producing microorganisms are mesophils with a maximum growth temperature of 30–40 °C. The maximum activity temperatures of a number of hydrogenases isolated from these microorganisms (Table 8) are the same.

For the thermophilic bacterium *Pseudomonas thermophila* the maximum temperatures of growth and hydrogenase activity also coincide at 50 °C [375]. However, for a number of purple bacteria (*R. rubrum*, *Rh. capsulata*, *Chr. vinosum*, *Th. roseopersicina*) which are the mesophils, the temperature optimum of hydrogenase activity is considerably higher than that of the culture growth (Table 8).

The temperature stability for *A. eutrophus Z1* [403] and *Sinechococcus elongatus* hydrogenases is close to the temperature optimum of their growth. However, for the hydrogenases of *Th. roseopersicina* and some other bacteria [351,336,398] the maximum of the temperature stability is much higher than the temperature optimum for their growth (Table 8).

To a certain extent, the stability of hydrogenases towards heating depends on the degree of purification and on the storage conditions of the enzymes. For example, the

Table 8. Thermostability and temperature optimum for hydrogenases of various microorganisms

Enzyme sources and purity	Electron donor (acceptor)	Reaction	Thermo-stability (°C)	$T_{opt.}$ (°C)	Ref.
Clostridium pasteurianum (cell-free extract)	MV	H_2 evolution	40	30	336)
C. butyricum (cell-free extract)	MV	H_2 evolution	40	30	336)
Desulfovibrio desulfuricans (partially purified)	MV	H_2 evolution	60	n.d.	398)
D. desulfuricans (partially purified)	MB	H_2 consumption	70	38	399)
D. desulfuricans (high purity)	MV	H_2 evolution	65	37	331)
D. vulgaris (homogeneous preparation)	MV	H_2 consumption	40	n.d.	400)
Methanobacterium thermo-autotrophicum (partially purified)	MV	H_2 consumption	65	n.d.	370)
Escherichia coli (highly purified)	MV	H_2 consumption	n.d.	32.5	364)
Citrobacter freundii (cell-free extract)	MV	H_2 evolution	60	n.d.	21)
Pseudomonas thermophila (membrane)	PMS+ DCPIP	H_2 consumption	50	50	375)
Prot. vulgaris (partially purified)	H_2	^{3}TO-H_2	50	n.d.	401)
Alcaligenes eutrophus H16 (partially purified)	NAD	H_2 consumption	n.d.	36	402)
A. eutrophus H16 (homogeneous, soluble)	NAD	H_2 consumption	n.d.	33	319)
A. eutrophus Z1 (cell-free extract)	MV	H_2 evolution	40	30	403)
	NADH		40	30	
A. eutrophus H16 (homogeneous, membrane)	MB	H_2 consumption	n.d.	32.5	364)
A. eutrophus Z1 (homogeneous, soluble)	MB	H_2 evolution	n.d.	30–37	404)
	NAD	H_2 consumption	n.d.		
Rhizobium japonicum (partially purified)	MB	H_2 consumption	70	60–70	405)
Rhodospirillum rubrum (cytoplasm)	MV	H_2 evolution	70	55	377)
(membrane)	MV	H_2 evolution	80	55	
Rhodopseudomonas capsulata (membrane)	BV	H_2 consumption	90	80	21)
(partially purified)	BV	H_2 consumption	78	70	380)
Thiocapsa roseopersicina (homogeneous)	MV	H_2 evolution	80	75	380)
Anabaena variabilis (cell suspension)	MV	H_2 evolution	60	50	18)
Spirulina platensis	MV	H_2 evolution	70	65	18)
Synechococcus elongatus (cell suspension)	MV	H_2 evolution	70	n.d.	18)
Chlorella vulgaris (cell suspension)	MV	H_2 evolution	65	60	300)

highly purified preparations of *D. desulfuricans* hydrogenase are completely inactivated after 5 min at 70 °C [339] while the partially purified enzyme under these conditions loses only 37% of its activity [398].

Of all the enzymes studied the hydrogenase of *Th. roseopersicina* is most stable to high temperatures [380]. The temperature optimum for this hydrogenase is 75 °C, heat inactivation of the enzyme starts at temperatures exceeding 80 °C.

The high ionic strength of the solutions markedly stabilizes *R. rubrum* hydrogenase [381]. But glycerol, which stabilized *Rh. capsulata* hydrogenase [528,529], does not influence the stability of *Th. roseopersicina* hydrogenase at elevated temperatures.

A current approach to increase the stability of hydrogenases towards oxygen inactivation and heat denaturation is their immobilization.

Immobilization of cells and hydrogenase preparations have been achieved for *Th. roseopersicina* [380,440], *C. butyricum* [325,456,542], *C. pasteurianum* [457], *D. vulgaris* [458], *A. eutrophus* [459], and *An. cylindrica* [270].

Immobilization of *Th. roseopersicina* cells in a polyacrylamide gel led to a 150–200 fold increase of the hydrogenase stability and an "activity period" of the enzyme of up to 6 month with retention of 80% of the initial activity [21].

The activity of *Th. roseopersicina* hydrogenase immobilized onto a semiconductor support (assayed via H_2 evolution rate with reduced MV) was 4times higher than that of the initial enzyme. This increase may be due to a specific bonding of the hydrogenase to the semiconductor matrix [460].

After immobilization of *C. pasteurianum* hydrogenase, possessing a low heat and oxygen stability [336,443], on glass beads, only 2–5% of the activity was retained. On the other hand, oxygen stability of the hydrogenase increased considerably [457,461]. A new technique of immobilization on ion exchange supports was suggested in which 50–60% of the starting catalytic activity of *C. pasteurianum* hydrogenase was retained [462]. A non-covalent bonding of the hydrogenase to the support increased its oxygen stability: 20–25fold in the case of DEAE-cellulose and 3000fold with polyethyleneimine-cellulose.

In the immobilization of *D. vulgaris* hydrogenase in a polyacrylamide gel the hydrogenase retains 20% of its initial activity after storage in air at 26 °C for a month whereas the starting enzyme under same conditions was completely inactivated after 10 days [458].

Thus, the simplest method of the stabilization of hydrogenases is their immobilization in a polyacrylamide gel. The exact immobilization technique, however, depends to a great extent on the intended applications of the enzyme and the conditions under which it is used.

4.1.6 Regulation of Biosynthesis and Activity

The synthesis of hydrogenases by different microorganisms depends on the availability of iron during their growth [21,29,33,35]. Furthermore, it was pointed out that some H_2-oxidizing bacteria, for example *A. eutrophus*, required nickel under autotrophic conditions of growth with H_2 [463,539,541]. Ammonium molybdate in the presence of sulfide represses hydrogenase synthesis in *Rum. albus* and *Rum. bromei*. It causes changes in the composition of substances which these microorganisms produce since the cells stop evolving H_2 [120]. Hydrogenase activity of *D. vulgaris* increases in the presence of EDTA. Furthermore, it depends on pH of the medium [334].

Some microorganisms, for example *N. opaca*, synthesize hydrogenase only under the conditions of autotrophic growth due to H_2 consumption. On the other hand, microorganisms are known to synthesize hydrogenase in the presence of some organic compounds regardless of the availability of H_2. These microorganisms include *A. eutrophus*[466)], some strains of *Paracoccus denitrificans*[465, 541)], *Methanosarcina barkeri*[540)], and some purple bacteria[182, 194, 464)]. The hydrogenase activity of some strains of *Par. denitrificans*, grown under heterotrophic conditions, is even higher than that of the cells grown under autotrophic conditions in the presence of H_2.

However, cases are known where the hydrogenase synthesis is competely repressed in the microorganisms grown on certain organic substrates even in the presence of H_2[466, 467, 541)]. Several microorganisms synthesize hydrogenases in media containing special organic substrates in the presence of H_2. Sometimes the presence of H_2 merely causes an increase of the hydrogenase activity. The nodule bacteria and a number of purple bacteria may serve as examples[468, 469)].

Incubation in a H_2-containing atmosphere sharply increases the hydrogenase activity of *An. cylindrica* and some other cyanobacteria[281, 282)]. From these data, the conclusion was drawn that molecular hydrogen may be essential for hydrogenase synthesis[470)].

Our studies on the purple bacterium *Rh. capsulata* have shown that the hydrogenase synthesis may be repressed by organic substances. However, the presence of H_2 is not a necessary although its availability slightly increases the hydrogenase activity in the cells.

Hydrogenase activity of *E. coli* and other H_2-producing enterobacteria sharply decreases when the cultures are grown in the presence of nitrates which these bacteria use as electron acceptors[373, 429)]. Facultative anaerobes generally reveal a higher hydrogenase activity or develop activity only under anaerobic conditions of growth[17, 21, 33, 109, 187, 429, 472)]. Even in the case of obligate aerobes may hydrogenase activity considerably depend on the O_2 content. For such hydrogen-oxidizing species as *Pseudomonas facilis* and *Alcaligenes rulandii*, grown under aerobic conditions, it was demonstrated that hydrogenase synthesis was feasible only when the O_2 concentration did not exceed 5% (v/v)[466)]. Oxygen markedly inhibits hydrogenase synthesis in such aerobes as *N. opaca*[473)], *Aquaspirillum autotrophicum*[474, 475)], *Xanthobacter autotrophicus*, *Azotobacter vinelandii*, and in the nodule bacteria[468, 469, 476, 478, 541)].

Hydrogenase activity of *An. cylindrica* and other N_2-fixing cyanobacteria increases several fold during anaerobic incubation. Experiments with chloramphenicol show that under these conditions hydrogenase synthesis occurs[55, 235, 247, 281, 282)].

As a rule, anaerobic preincubation is essential for the development of hydrogenase activity in algae. In the course of preincubation not only the activation of hydrogenases but possibly also their synthesis occurs[18, 36–48, 222)].

From all the available data it follows that some microorganisms are able to synthesize hydrogenases irrespective of the composition of the growth medium although a number of factors influence the amounts of these enzymes. The synthesis of hydrogenases in several microorganisms depends on the H_2 availability or is derepressed in the absence of organic substrates. However, not all organic substances possess the ability to repress hydrogenase synthesis[466, 475)].

In general, hydrogenase activity in vivo depends on the same factors influencing the enzyme synthesis. From the data discussed above it becomes obvious that the hydrogenase-catalyzed H_2 evolution usually proceeds under anaerobic conditions. It is generally determined by oxygen inhibition of hydrogenases; sometimes, the effect of O_2 is irreversible.

Oxygen sensitivity of the hydrogenases which catalyze H_2 oxidation by O_2 is lower. But even in this case, high O_2 concentrations may lead to enzyme inactivation for example in the case of nodular bacteria [468,478].

Furthermore, in certain cases, the activity of hydrogenase H_2 evolution may be suppressed in the presence of high levels of hydrogen. This is probably related to the redox potentials of the donors interacting with the hydrogenases [31,33,35,41,165,309].

To conclude, the factor determining the hydrogenase activity of various microorganisms (demonstrated as the ability to evolve and to consume molecular hydrogen) have been elucidated. The main factors inhibiting hydrogenase activity of cells have been established. Nevertheless, the actual mechanisms regulating the synthesis and the activity of these enzymes are not yet understood.

4.2 Nitrogenase

Nitrogenase (EC 1.7.99.2) catalyzes the reduction of molecular nitrogen to ammonia:

$$N_2 + 6e^- + 6H^+ + nATP \rightarrow 2NH_3 + nADP + nP_i$$

This enzyme can also catalyze the reduction of a number of other compounds with triple bonds [479–481] and the irreversible energy-dependent reduction of protons to molecular hydrogen [132,175,334,482]:

$$2H^+ + 2e^- + ATP \rightarrow H_2 + ADP + P_i$$

Nitrogenase is also involved in a specific exchange reaction which proceeds in the presence of N_2 [23,483,484]. Since the ability to catalyze H_2 evolution is probably the common property of the nitrogenases of all microorganisms, it may be assumed that all N_2-fixing species may, in principle, evolve hydrogen, due to the function of this enzyme.

Recently, several reviews on the properties of nitrogenase have been published [23,24,132,481,483–485,530,531].

Nitrogenases of all microorganisms consist of two components, a molybdenum iron-containing protein (MoFe-protein, component I or molybdoferredoxin) and an iron-containing protein (Fe-protein, component II or azoferredoxin). Sometimes, the Fe-protein is called nitrogenase reductase and the MoFe-protein nitrogenase itself [175,486,487].

The MoFe-protein is a high molecular weight complex ($200–300 \times 10^3$) and consists of four subunits (Table 9).

In comparison with the MoFe-protein the molecular weight of the Fe-protein is lower ($M \sim 51–73 \times 10^3$ (Table 9)). It consists of two subunits with the same molecular weight ($27.5–34.6 \times 10^3$).

The MoFe-protein of nitrogenases contains 18–38 atoms of non-heme iron, 18–38 atoms of acid-labile sulfur, and 1–2 atoms of molybdenum per molecule [132,

Table 9. Properties of nitrogenase isolated from various bacteria

Enzyme sources	$M_r \times 10^3$	Subunits $\times 10^3$	Content			EPR-signals (g-factors)	Specific activity, nmol C_2H_4 min^{-1} mg^{-1} protein	Ref.
			Mo	Fe	S^{2-}			
MoFe-protein								
Clostridium pasteurianum	200–220	2 × 50; 2 × 59.8	2	18–24	18–24	4.23; 3.77; 2.01	2250	[406]
Azotobacter vinelandii	270–300	4 × 70	2	34–38	26–28	4.32; 3.65; 2.01	1400	[407]
Ab. chroococcum	222	2 types	2	22 ± 2	20 ± 2	4.29; 3.65; 2.01	2000	[408]
Bacillus polymyxa	215	n.d.	2	32.8	21.0	4.37; 2.53; 2.01	2750	[409]
Klebsiella pneumoniae	218	4 × 51	2	27–33	n.d.	4.32; 3.63; 2.01	2150	[410]
Xanthobacter autotrophicus	237.3	2 × 57; 2 × 61	2.2	23.1	20.01	n.d.	1070	[411]
Rhizobium japonicum	200	n.d.	1.3	28.8	28.2	4.17; 3.75; 2.03	900	[412]
Rz. lupini	104	n.d.	1	18–20	n.d.	n.d.	704	[413]
Rhodospirillum rubrum	215–230	2 × 56; 2 × 58	2	20–30	19–22	4.34; 3.65; 2.01	920–1718	[414]
Chromatium vinosum	160	2 types	1.4	19	15	4.30; 3.68; 2.01	1600	[415]
Anabaena cylindrica	223	2 × 52.8; 2 × 55	2	20 ± 2	20 ± 2	n.d.	1200	[416]
Fe-protein								
C. pasteurianum	56	2 × 27.5	—	4.0	4.0	2.05; 1.94; 1.88	2708	[406]
Ab. vinelandii	64	2 × 33	—	3.5–4.0	2.8–4.0	2.05; 1.94; 1.88	1100	[407]
Ab. chroococcum	65	n.d.	—	4.0	3.9	2.05; 1.94; 1.86	2000	[408]
Bac. polymyxa	55.5	n.d.	—	3.2	3.6	n.d.	2521	[409]
K. pneumoniae	62	2 × 34.6	—	4.0	3.8	2.05; 1.94; 1.87	980	[410]
X. autotrophicus	73	n.d.	—	3.8	2.4	n.d.	1260	[411]
Rz. japonicum	51	n.d.	—	1.0	n.d.	n.d.	n.d.	[412]
Rz. lupini	65	n.d.	—	3.2	n.d.	n.d.	434	[413]
R. rubrum	60–65	2 × 31.5	—	3.8–4.5	2.8–5.0	n.d.	1260	[417]
An. cylindrica	60	n.d.	—	n.d.	n.d.	n.d.	300	[416]

[175,442,488]. About half the iron and molybdenum is bound to a lowpotential cofactor: these metals may be detached from the MoFe-protein by acid treatment [481,489,490].

Four atoms of non-heme iron and four atoms of acid-labile sulfur (S^{2-}) are found in a dimer of the Fe-protein (Table 9).

4.2.1 Mechanism of Catalysis and H_2 Evolution

All nitrogenase-catalyzed reactions demand the participation of both components of the nitrogenase. A recombination of the isolated components of the nitrogenases from different organisms is possible [408,409].

For nitrogenase to function, energy in the form of ATP and Mg ions are necessary. According to the available data, 12–30 mol of ATP are required for the reduction of each N_2 molecule to NH_3 [132,488]. This enables a transfer of electrons from the Fe-protein to the MoFe-protein [175,479]. With the exception of the exchange reaction, in all cases nitrogenase activity also depends on the presence of a low-potential reductant [132,491].

For the activation of *R. rubrum* Fe-protein [326,492–494] as well as perhaps that of *Rh. palustris* [207], *Rh. capsulata* [202], and *Az. lipoferum* [409] a specific membrane-bound enzyme is necessary under certain conditions. The molecular weight of this enzyme from *R. rubrum* which is stabilized in the presence of Mn is about 20500 [202,284].

The affinity of nitrogenase to N_2 is higher than its affinity to the other substrates. The Michaelis constant for N_2 lies in the range 0.03 to 0.1×10^{-3} M whereas for other substrates this constant is only 1.2×10^{-2}–3.6×10^{-4} M [483,485].

Nitrogenase catalyzes the formation of HD from D_2 and of H_2O from H_2 and D_2O [183,181]. The formation of HD is stimulated by N_2. It is probably formed by exchange reaction of D_2 with the hydrogen of the intermediate products of the reduction (bound to the enzyme) of diimide or hydrazine with N_2. It is assumed that such "exchange" reaction (H^+/D_2O) is carried out by nitrogenase (E) in the following way [23,484]:

$$E + N_2 \longrightarrow E:N_2 \xrightarrow[2H^+]{nATP,\, 2e^-} E:N_2\,H_2 \xrightarrow{+D_2} E:N_2 + 2HD \longrightarrow E + N_2$$

Nitrogenase-catalyzed H_2 evolution was first observed when dithionite, ATP, and Mg^{2+} were added to enzyme preparations from *Ab. vinelandii* and *R. rubrum* [482]. The most intense evolution of H_2 catalyzed by nitrogenase both in vitro and in vivo occurs in the absence of N_2 which is a competitive inhibitor of this process [485,495]. Nevertheless, nitrogenase-catalyzed H_2 evolution may proceed in the presence of N_2, both in whole cells and in cell-free preparations.

The ratio of the rates of the reduction of H^+ and N_2 varies considerably for nitrogenases of different N_2-fixing organisms [75,479]. But even for pN_2 values which are sufficiently high for nitrogen fixation, about 25% of energy may be spent on H_2 evolution by nitrogenase [132,479]. Nitrogenase-catalyzed H_2 evolution by *Anabaena* sp. *7120* heterocysts decreases at 68% of N_2 in the atmosphere. However, in the presence of 18% acetylene this evolution is entirely suppressed. From this it follows that acetylene and N_2 inhibit H_2 evolution and affect nitrogenase in different ways [479].

The rates of both H_2 evolution and N_2 reduction, catalyzed by nitrogenase, depend on the ratios of the MoFe-protein to the Fe-protein [23,411] and of ATP to ADP [496].

It was demonstrated for *X. autotrophicus* nitrogenase preparations that an increase in the MoFe-protein content causes a rise of H_2 evolution in the N_2 atmosphere [411].

H_2 evolution is probably related to the normal mechanism of nitrogenase function in the course of which six electrons are spent for the reduction of N_2 and at least two for the production of H_2 [23,442,479,484]. Accordingly, the process of nitrogenase catalysis may be described by the following equation [76]

$$N_2 + 8e^- + 8\,H^+ + nATP \rightarrow 2\,NH_3 + H_2 + nADP + nP_i$$

The binding sites of N_2 and H_2 on a nitrogenase are apparently different. Carbon monoxide and carbamyl phosphate do not affect H_2 evolution while they inhibit N_2 demonstrated by means of an artificial system containing vanadium as a catalyst [497].

This mechanism is in line with the view that nitrogenase can catalyze the transfer of four rather than of two electrons [481,497]. The feasibility of such process was reduction and all other nitrogenase-catalyzed reactions [483].

It is assumed [481] that H_2 production is catalyzed by a nitrogenase form with a distorted structure of its binuclear center. Such a form of nitrogenase is able to evolve H_2 and to catalyze the slow exchange of deuterium, as well as to reduce compounds requiring at least one or two electrons. But it is incapable of N_2 reduction.

4.2.2 Regulation of Biosynthesis and Activity

For the synthesis of nitrogenase to take place, it is necessary to provide the growing bacteria with iron and molybdenum since these elements are included into the enzyme molecule [132,200,201]. In the case of some purple bacteria the presence of manganese is also important. Manganese is essential for the action of the nitrogenase-activating enzyme [202].

Of considerable importance for nitrogenase biosynthesis by various bacteria is the oxygen content. Although *Azotobacter* and several other microorganisms are able to grow under aerobic conditions due to nitrogen assimilation, for nitrogenase synthesis lower pO_2 values are favourable [23,24,484,488,532]. A number of N_2-fixing organisms that require oxygen for their growth synthesize nitrogenases only under microaerobic or anaerobic conditions. Examples are *Az. lipoferum* [498] and certain cyanobacteria [23,34,174,219,231,499]. Nitrogenase synthesis by *K. pneumoniae*, which is a facultative anaerobe, demands strictly anaerobic conditions [23,484].

Considerable concentrations of NH_4^+ exert a repressable effect on nitrogenase synthesis by various bacteria [24,53,132,174,187,488].

According to a number of data on *K. pneumoniae*, *Ab. vinelandii*, the purple bacteria, and some other microorganisms, repression and derepression of nitrogenase synthesis depends on the state of glutamine synthetase, this enzyme plays an important role in the metabolism of ammonium, especially when its concentration is low. The following data furnish evidence of the role of glutamine synthetase in the regulation of nitrogenase synthesis. Firstly, it was shown that the *K. pneumoniae* mutants and the mutants of certain purple bacteria with derepressed nitrogenase are ineffective in glutamine synthetase. Secondly, in the presence of methionine sulfone, methionine sulfoximine and other compounds which are bound to glutamine synthetase and thus prevent it from its function, a number of bacteria are able to synthesize nitrogenase in the presence of NH_4^+ [24, 132, 232, 484, 488, 500].

It is supposed that glutamine synthetase acts as a positive effector of nitrogenase synthesis when it is present in the active deadenylated form which is the case when there is a deficiency of ammonium. However, regulation of the activity of glutamine synthetase itself (through adenylation-deadenylation) does not take place in all N_2-fixing organisms. For example, it was not discovered in the Cyanobacteria [24]. It cannot be ruled out that regulation of nitrogenase synthesis is carried out not by glutamine synthetase but by glutamine and other products synthesized from ammonium as the result of its functioning [24, 132, 500–502].

Along with the regulation of synthesis, the activity of nitrogenase may change, depending on the conditions of the N_2-fixing bacteria.

ADP shows an inhibitory influence on nitrogenase. Thus, an increase of the ADP/ATP ratio in the cells may lead to a sharp decrease in nitrogenase activity [484, 488].

Oxygen is one of the most significant factors affecting not only the synthesis but also the activity of the nitrogenases of various microorganisms. Furthermore, oxygen may cause inactivation of nitrogenase which is generally irreversible. Therefore, microorganisms are able to perform N_2 fixation and to evolve H_2, due to nitrogenase activity only under anaerobic conditions. Comparatively few bacteria may perform these processes under microaerobic and even less under aerobic conditions [53, 132, 231, 484, 503, 533]. Nevertheless, from the latter case it does not follow that nitrogenases of these microbes are more stable to oxygen than that of anaerobes. Nitrogenase activity in the presence of O_2 is explained by the ability of these organisms to protect the enzyme against the inactivating effect of O_2 [23, 24, 484, 488, 516, 534].

In the case of *Azotobacter* and probably some other N_2-fixing organisms, a high respiration rate is of considerable importance [23, 534]. Due to this high rate there is not sufficient time available for O_2 to inactivate nitrogenase (respiratory protection). Besides, *Azotobacter* and several other N_2-fixing organisms possess slime capsules which may also be important for the protection of nitrogenase against oxygen. Location of nitrogenase in cells may be significant for its protection against oxygen (conformational protection).

Nitrogenase of the filamentous cyanobacteria may function in vivo under aerobic conditions. As mentioned above, it is located in heterocysts with a very thick envelope. Moreover, in contrast to vegetative cells, heterocysts do not evolve O_2, but are actively involved in the consumption of the latter. So in this case, a further way of protecting nitrogenase against O_2 is effective, namely the special separation of N_2 fixation from photosynthesis which is accompanied by O_2 evolution.

In the nodules of leguminous plants leghemoglobin apparently takes part in O_2 protection of the nitrogenase of bacteroides and lowers the concentration of oxygen. It is also suggested that H_2 production by N_2-fixing organisms might be a way of protecting nitrogenase against high oxygen concentrations [23, 83, 131].

However, the mechanism of nitrogenase inactivation by O_2 remains unclear. Probably, it is related to certain reaction products of oxygen, e.g. to O_2^*, O_2^- and H_2O_2 [448].

Apart from O_2, the inhibition of the nitrogenase activity of the purple bacteria and cyanobacterium *An. cylindrica* is caused by ammonium [477, 535]. However, the effect of ammonium is reversible, and when this compound is assimilated by the cells, nitrogenase becomes active again [187, 191, 194, 207–209, 464, 504–506, 535]. There are also data on the inhibition of H_2 evolution by ammonium in *Rhizobium leguminosarum* bacteroids [536]. The inhibiting effect of NH_4^+ on the nitrogenase of *R. rubrum*, *Rh. palustris* and *Rh. capsulata* depends on the growth conditions of the latter. Ammonium does not effect nitrogenase of the cells grown under nitrogen-limiting conditions [21, 535]. It was supposed that glutamine synthetase takes part not only in the regulation of nitrogenase synthesis but also in the regulation of nitrogenase activity as well as in some bacteria [207, 208, 507].

According to recent results, inhibition of nitrogenase activity of *R. rubrum* in the presence of NH_4^+ is determined by the conversion of this enzyme to an inactive regulated form. Simultaneously, the conversion of glutamine synthetase to the inactive adenylated form is observed [494]. However, the nitrogenase activity is possibly not regulated by the direct participation of glutamine synthetase but affected by glutamine or other products formed as a result of this activity of the enzyme [500, 501, 505, 508]. Recent data prove that glutamine synthetase is not directly involved in the regulation of *Rh. palustris* and *Rh. sphaeroides* nitrogenase [522].

Along with ammonium, the activity of nitrogenase of *An. cylindrica* and of several purple bacteria is inhibited in the presence of glutamine, asparagine, urea, NO_3^-, and NO_2^- [187, 207, 208, 464, 504, 506, 509]. However, these compounds do not affect all N_2-fixing organisms [484].

Meanwhile, it is known that nitrogenase activity, assayed *via* the reduction of N_2 or C_2H_2, is inhibited in different bacteria by carbamyl phosphate and by CO. However, H_2 evolution, is not influenced by these substances [484].

It should be pointed out that the capability of microorganisms for N_2 fixation and for nitrogenase- or hydrogenase-mediated H_2 evolution may depend on the activity of other cellular components involved in these processes. Oxygen, for instance, may not only cause the inactivation of nitrogenase and hydrogenase but, in certain microorganisms also oxidize the electron carriers interacting with the enzymes. As a result, H_2 formation by the cells ceases [448].

4.3 Interrelationship between Hydrogenase and Nitrogenase

An interrelationship between hydrogenase and nitrogenase was established after the discovery of the nitrogenase-mediated H_2 evolution [53, 131, 187, 210, 220, 265, 464, 510]. For a number of *Rhizobium* strains it was shown that the energy spent for H_2 evolution, catalyzed by nitrogenase, amounts to 25–60% even in the presence of N_2 [75, 131, 133]. However, not all strains of *Rhizobium* are so active in hydrogen evolution. For a number of nodular bacteria, H_2 produced with nitrogenase participation may be rapidly consumed by the same microorganisms, due to hydrogenase activity, i.e. a so-called H_2 recyclization process takes place [511, 537].

Therefore, in some cases, molecular hydrogen evolution by N_2-fixing organisms can be detected (or it is markedly increased) with the inhibition of the activity of H_2-consuming hydrogenases. This is achieved often by adding CO or CO and acetylene to the cell suspensions [82, 83].

The available data confirm the ability of H_2 recyclization not only in the case of the nodular bacteria [76, 131, 135, 512, 513] but also for *Azotobacter* [83, 514], *Az. brasiliense* [84], *Az. lipoferum* [538], *Xanthobacter autotrophicus* [135], for a number of cyanobacteria [265, 266, 283], and some purple bacteria [53, 196, 203, 214, 464, 510].

Thus, hydrogen recyclization is characteristic of several N_2-fixing organisms. The few exceptions are some bacterial strains which do not produce hydrogenase or the bacteria which possess a weakly developed H_2-consuming system. Hydrogen recyclization in various microorganisms may be of different importance but in all cases it activates the reaction [21, 53, 54, 131, 512].

Consumption of hydrogen, produced by nitrogenase, prevents this enzyme from being inhibited by H_2. In the case of aerobic N_2-fixing organisms the recycled H_2 may be oxidized by O_2. This is probably important for nitrogenase protection against the inhibitory action of O_2. Moreover, oxidation of H_2 with oxygen as the terminal electron acceptor may provide microorganisms with additional energy. This process was confirmed for *Azotobacter* [83], the nodular bacteria [131] and for the cyanobacteria [54, 55, 515]. Some microorganisms utilize the recycled hydrogen as an electron donor of nitrogenase [21, 83].

There are data which suggest that nitrogenase and hydrogenase synthesis in certain microorganisms, for example in *Ab. chroococcum* and in others, proceeds in a coordinated manner. However, according to other data such a phenomenon is not always observed [21, 187, 210, 383, 464, 510].

Thus, from the available data the conclusion may be drawn that the coordinated activity of nitrogenase and hydrogenase is physiologically important for microorganisms, thus increasing their N_2-fixing activity. However, when microorganisms are selected for H_2 production and if the desired process is nitrogenase-catalyzed, the presence of the active H_2-consuming hydrogenase is undesirable. In order to anticipate hydrogen recyclization by such microorganisms, inhibitors preventing H_2 oxidation may be also used [21, 54, 222].

4.4 Applications of Hydrogenases and Nitrogenase

Investigations of the properties of hydrogenases and nitrogenase are of interest not only for a better understanding of H_2 metabolism and N_2 fixation in various microorganisms but also for the elucidation of the mechanisms of fermentative catalyses. These enzymes may find practical applications in different processes.

For example, *A. eutrophus* hydrogenase, which reduces NAD directly by H_2, may be used as an NADH-regenerator in the synthesis of a range of organic compounds [517]. Hydrogenases may also be applied in electrochemical systems. Bioelectrocatalyses employing hydrogenases may be used as a basis of the design of fuel cells of electrochemical transducers of energy [158, 518]. Furthermore, there arises the possibility of using hydrogenases in electrosynthesis and for analytical purposes [325]. A method is worked out for the manufacture of model electrodes utilizing the immobilized hydrogenase for electrochemical H_2 oxidation [518].

The problem of coupling hydrogenase and nitrogenase functions with the process of photosynthesis has attracted great attention; the ultimate aim is the production of molecular hydrogen [17, 21, 222, 303, 304, 519]. Especially, much work was done to investigate H_2 production in systems containing chloroplasts and hydrogenases based on water biophotolysis. Such systems usually also contain ferredoxin or other electron carriers with similar redox potentials [21,420–422,471] as well as the components ensuring the removal of oxygen or its reduction products which inhibit the production of hydrogen. These systems may be composed of glucose and glucose oxidase, ethanol and catalase, or contain other compounds [418,421,424,425].

In a comparatively short period of time, a considerable increase of the rate of H_2 evolution has been achieved by using such systems. The stability of the systems concerned was also raised (Table 10).

In first experiments with chloroplast-containing systems, the obtained rate of H_2 evolution was 1–2 μmol h^{-1} mg^{-1} of chlorophyll and the process continued for 15 to 20 min [418]. According to the latest data this process goes on for six or more hours, the rate of H_2 evolution increasing to 100 μmol h^{-1} mg^{-1} of chlorophyll [425,519]. The maximum efficiency of the energy conversion in the system containing chloroplast and hydrogenases was calculated to be about 25% [59].

Nevertheless, the action of such systems remains limited and of low efficiency. It is determined by a number of factors such as the activity and stability of chloroplasts, the property of the hydrogenase and electron carriers used, the temperature and ionic strength of the medium, the light intensity, etc.

Table 10. Photoevolution of H_2 by cell-free systems containing chloroplasts

Basic components of the system	Electron donor	H_2, μmol h^{-1} mg^{-1} chlorophyll	Functioning time (h)	Ref.
Chp^b + Fd + H_2-ase *C. kluyveri*	H_2O	1–2	0.25	418)
	H_2O	12–14*	?	
	ascorbate + DCPIP	15.4	?	
Chp^c + H_2-ase *Chl. reinhardii*	DTT	4.3	1.0	291)
Chp^d + MV + H_2-ase *Th. roseopersicina*	NADH	8.2	?	419)
Chp^b + Fd + H_2-ase *C. pasteurianum*	H_2O	9.0–15.5^a	3.0–6.0	420)
Chp^b (immobilized) + Fd + H_2-ase *C. pasteurianum*	H_2O	11.5^a	3.0	420)
Chp^b + Fd + H_2-ase *E. coli*	H_2O	2.8	3.0	420)
Chp^b + Fd + H_2-ase *C. pasteurianum*	H_2O	7.0–13.0^a	2.0–6.0	421)
Chp^e + Fd + H_2-ase *Th. roseopersicina*	H_2O	5.0–10.0^a	3.0–6.0	422)
Chp^b + Fd(MV) + H_2-ase *C. pasteurianum*	TMPD	50–125	0.17	423)
Chp^b + Fd + H_2-ase (from various sources)	H_2O	20–40^a	4.0	424)
Chp^b + Fd + H_2-ase (from various sources)	H_2O	30–50^a	2.0–6.0	425)
Chp^f + Fd + H_2-ase *C. pasteurianum*	H_2O	94^a	2.0–4.0	425)
Chp^d + MV + H_2-ase *C. butyricum*	ascorbate + DTT	350–400	?	426)
Chp^b + Fld *Anacystis nidulans* + H_2-ase *C. pasteurianum*	H_2O	27.2	?	427)

Note. a: The experiments were carried out in the presence of O_2 and H_2O_2 scavengers, the rest variants, in the absence of scavengers. The sources of chloroplasts (Chp) were: b — spinach; c — *Chlamydomonas reinhardii*; d — pea; e — tobacco; f — *Chenopodium album*

Apart from the chloroplast-containing systems, hydrogenases instead of PtO_2 are used in the catalysis of the last stage of H_2 production in systems containing chlorophyll [303], nicotinamide adenine dinucleotides [419] or dyes [10,17,21,304,519,520].

5 Conclusion

Extensive investigations conducted in the recent years in different laboratories revealed a number of new chemotrophic and phototrophic microorganisms capable of H_2 production. Some of these microorganisms evolve considerable amounts of H_2 at a rather high rate. However, the utilization of the chemotrophic microorganisms for the production of hydrogen is less advantageous than the microbiological production of methane based on mixed cultures incorporating usually H_2-producing species.

More promising for hydrogen production are phototrophs. Some of them are able to form H_2 with a rather high efficiency of solar energy conversion.

Much attention has especially been paid to the purple bacteria and cyanobacteria. The application of the purple bacteria to hydrogen production mainly depends on the availability of cheap oxidizable organic substrates and on their resources. Still more attractive producers of H_2 are certain cyanobacteria since the evolution of hydrogen in this case is linked with the biophotolysis of water. However, before these organisms can be utilized for the production of H_2 much work has to be done in selecting the strains with the highest activity (especially among the marine species), and in determining the optimal conditions for the realization and stabilization of this process.

Substantial progress was also made in the studies of hydrogenases and nitrogenases catalyzing H_2 evolution in various microorganisms. The utilization of the hydrogenases of certain microorganisms for practical purposes, such as bioelectrocatalysis and regeneration of reduced cofactors, provides realistic prospects for the future.

Moreover, hydrogenases are being used for H_2 production in cell-free systems, which may serve as prototypes of the artificial systems of H_2 production, based on water biophotolysis. Since in the majority of phototrophic microorganisms hydrogen evolution is nitrogenase-mediated, these microorganisms may be used not only for H_2 production but also for the synthesis of ammonia from N_2, based on the bioconversion of solar energy.

6 Abbreviations

MV	methyl viologen
BV	benzyl viologen
MB	methyl blue
Fd	ferredoxin
Fld	flavodoxin
NAD(P)	nicotinamide adenine dinucleotide (phosphate)
DCPIP	dichlorophenol indophenol

PMS	phenazine methosulfate
DTT	dithiothreitol
TMPD	N,N,N′,N′-tetramethyl-p-phenylenediamine
CoA	coenzyme A
DCMU	3-(3,4-dichlorphenyl)-1,1-dimethylurea
POD	peroxide dismutase
H_2-ase	hydrogenase
EDTA	ethylenediamine tetraacetate

7 References

1. Gregory, D. P., Pangborn, I. B.: Ann. Rev. Energy *1*, 279 (1976)
2. Weissman, J. C., Benemann, J. R.: Appl. Environ. Microbiol. *33*, 123 (1977)
3. Holmström, B.: In: Solar Energy-Photochemical Conversion and Storage (Classon, S., Engström, L. eds.), p. 1, Stockholm: Liber Tryck 1977
4. Higgins, I. J., Hill, H. A. O.: In: Microbial Technology Current State. Future Prospects, p. 359. London, New York: Cambridge Univ. Press 1978
5. Hall, D. O.: Solar Energy *22*, 307 (1979)
6. Malik, K. A.: Process Biochem. *13*, 4 (1979)
7. Miyamoto, K., Miura, Y.: ibid: *15*, 23 (1980)
8. Weaver, P. F., Lien, S., Seibert, M.: Solar Energy *24*, 3 (1980)
9. Bachofen, B.: Experientia *36*, 1429 (1980)
10. Adams, M. W. M., Mortenson, L. E., Chen, J.-S.: Biochim. Biophys. Acta *594*, 105 (1980)
11. Zeikus, J. G.: Bacteriol. Revs. *41*, 514 (1977)
12. Benemann, J. R. et al.: Nature *268*, 5615 (1977)
13. Bryant, M. P.: J. Anim. Sci. *48*, 193 (1979)
14. Wolfe, R. S.: Antonie van Leeuwenhoek. J. Microbiol. Serol. *45*, 353 (1979)
15. Wolfe, R. S.: In: Microbial Biochemistry (Quale, R. S. ed.), p. 363. Baltimore: Univ. Park Press 1979
16. Bonch-Osmolovskaja, E. A.: Usp. Microbiol. *14*, 106 (1979)
17. Krasna, A. I.: Enzyme and Microbiol. Technol. *1*, 165 (1979)
18. Gogotov, I. N.: Usp. Mikrobiol. *14*, 3 (1979)
19. Zeikus, J. G.: Ann. Microbiol. *34*, 423 (1980)
20. Wiegel, J.: Experientia *36*, 12 (1980)
21. Kondratieva, E. N., Gogotov, I. N.: Molecular Hydrogen in Microbial Metabolism. Moscow: Nauka 1981
22. Hardy, R. W. F., Silver, W. S. (eds.): A Treatise on Dinitrogen Fixation. New York, London, Sydney, Toronto: Wilsey — Intersci. Publ. 1977
23. Robson, R. L., Postgate, J. R.: Ann. Rev. Microbiol. *34*, 183 (1980)
24. Stewart, W. D. P.: ibid. *34*, 497 (1980)
25. Stewart, W. D. P., Gallon, J. R. (eds.): Nitrogen Fixation. London, New York: Academic Press 1980
26. Zo Bell, C. E.: Bull. Amer. Assoc. Pet. Geologist. *31*, 1709 (1947)
27. Gest, H.: Bacteriol. Revs. *18*, 43 (1954)
28. Grey, C. T., Gest, H.: Science *148*, 186 (1965)
29. Quadri, S. M. H., Kaikobad, Y. M.: Pakistan J. Sci. *21*, 85 (1969)
30. Schlegel, H. G.: J. Geophys. Tellus *26*, 11 (1974)
31. Mortenson, L. E., Chen, J. S.: In: Microbial Iron Metabolism (Neilands, J. B. ed.), p. 231. New York: Academic Press 1974
32. Cole, J. A.: In: Adv. Microbial. Physiol. *14*, 1 (1976)
33. Kondratieva, E. N., Gogotov, I. N.: Izvestija Acad. Nauk SSSR, seria biol. *1*, 69 (1976)
34. Smith, G. D.: Search *9*, 209 (1978)
35. Zajic, J. E., Kosaric, N., Brossean, J. D.: Adv. Biochem. Eng. *9*, 57 (1978)

36. Schlegel, H. G., Schneider, K.: In: Hydrogenases: Their Catalytic Activity, Structure and Function (Schlegel, H. G., Schneider, K. eds.), p. 15. Göttingen: E. Goltze 1978
37. Kondratieva, E. N.: In: The Role of Microorganisms in the Hydrogen Cyclus in Nature (Zavarin, G. A. ed.), p. 131. Moscow: Nauka 1979
38. Barker, H. A.: In: The Bacteria. A Treatise on Structure and Function (Gunsalus, I. C., Stanier, R. Y. eds.), Vol. II, p. 156. New York, London: Academic Press 1961
39. Wood, U.: In: The Bacteria. A Treatise on Structure and Function (Gunsalus, I. C., Stanier, R. Y. eds.), Vol. II, p. 63. New York, London: Academic Press 1961
40. Thauer, R. K., Jungermann, K., Decker, K.: Bacteriol. Revs. *41*, 100 (1977)
41. Gottschalk, G., Andreesen, J.: In: Microbial Biochem. (Quayle, J. R. ed.), p. 85. Baltimore: Univ. Park Press 1979
42. Gest, H., Kamen, M. D.: In: Handb. der Pflanzenphys. Vol. II, p. 568. Berlin: Springer 1960
43. Gest, H.: Adv. Microb. Physiol. *7*, 243 (1972)
44. Ormerod, I. G., Gest, H.: Bacteriol. Revs. *26*, 51 (1962)
45. Pfennig, N.: Ann. Rev. Microbiol. *21*, 286 (1967)
46. Kessler, E.: In: Algae Physiology and Biochemistry (Stewart, W. D. P. ed.), p. 456. Oxford: Blackwell 1974
47. Kessler, E.: In: Microbial Production and Utilization of Gases (Schlegel, H. G., Gottschalk, G., Pfennig, N. eds.), p. 247. Göttingen: E. Goltze 1976
48. Kessler, E.: In: Hydrogenases: Their Catalytic Activity, Structure and Function (Schlegel, H. G., Schneider, K. eds.), p. 415. Göttingen: E. Goltze 1978
49. Lien, S., San Pietro, A.: An Inquiry into Biophotolysis of Water to Produce Hydrogen. Bloomington: Dept. Plant. Sci., Indiana Univ. 1975
50. Ochepkov, V. P., Krasnovsky, A. A.: Izvestija Akad. Nauk SSSR, seria biol. *1*, 87 (1976)
51. Kondratieva, E. N.: In: Microbiological Energy Conversion (Schlegel, H., Barnea, J. eds.), p. 205. Göttingen: E. Goltze 1976
52. Kondratieva, E. N.: Prikladnaja Biochimia i Microbiologia *14*, 805 (1978)
53. Meyer, V., Kelley, B. C., Vignais, P. M.: Biochemie *60*, 245 (1978)
54. Bothe, H., Distler, E., Eisbrenner, G.: ibid. *60*, 277 (1978)
55. Bothe, H., Eisbrenner, G.: In: Hydrogenases: Their Catalytic Activity, Structure and Function (Schlegel, H., Schneider, K. eds.), p. 353. Göttingen: E. Goltze 1978
56. Gibbs, M. et al. (eds.): Proc. of the Workshop on Biosolar Conversion. NSF/RANN Report 1973
57. Beck, R. W. (ed.): Workshop on Solar Energy for Nitrogen Fixation and Hydrogen Production. Knoxville: Univ. Tennessee 1975
58. Schlegel, H., Gottschalk, G., Pfennig, N. (eds.): Microbial Production and Utilization of Gases. Göttingen: E. Goltze 1976
59. Schlegel, H., Barnea, J. (eds.): Microbiological Energy Conversion. Göttingen: E. Goltze 1976
60. Buvet, R., Allen, M. J., Massue, J. P. (eds.): Living Systems as Energy Converters. Amsterdam: Elsevier/North-Holland Biomed. Press 1977
61. Europ. Sem. on Biological Solar Energy Conversion Systems. Available from UK-ISES, 21 Albemarle st. London 1977
62. Mitsui, A. et al. (eds.): Biological Solar Energy Conversion. New York: Academic Press 1977
63. Schlegel, H. G., Schneider, K. (eds.): Hydrogenases: Their Catalytic Activity, Structure and Function. Göttingen: E. Goltze 1978
64. Barber, J. (ed.): Photosynthesis in Relation to Model Systems. Amsterdam: Elsevier/North-Holland Biomed. Press 1979
65. Zeikus, J. G., Hegge, P. W., Anderson, M. A.: Arch. Microbiol. *122*, 41 (1979)
66. Wiegel, J. G., Ljungdahl, L. G.: ibid. *128*, 343 (1981)
67. Ben-Bassat, A., Zeikus, J. G.: ibid. *128*, 365 (1981)
68. Boone, D. R., Bryant, M. P.: Appl. Environ. Microbiol. *40*, 626 (1980)
69. Braun, K., Gottschalk, G.: Arch. Microbiol. *128*, 294 (1981)
70. Bergey's Manual of Determinative Bacteriology 8th Baltimore: The Williams Wilkins Co. 1974

71. Stanier, R. J., Adelberg, E. A., Ingraham, J.: The Microbial World. New Jersey: Prentice-Hall 1976
72. Laanbrock, H. J., Stal, L. J., Veldkamp, H.: Arch. Microbiol. *119*, 99 (1978)
73. Vollbrecht, D. et al.: Eur. J. Appl. Microbiol. *7*, 267 (1979)
74. Kavamura, S., Wilkinson, D. F.: In: Abstracts of 2nd Int. Symp. Growth of Microorganisms on C_1-compounds, p. 65. Pushchino: NCBI 1977
75. Schubert, K. R., Evans, H. J.: Proc. Nat. Acad. Sci. USA *73*, 1207 (1976)
76. Evans, H. J. et al.: In: Hydrogenases: Their Catalytic Activity, Structure and Function (Schlegel, H. G., Schneider, K. eds.), p. 287. Göttingen: E. Goltze 1978
77. Ruiz-Argüeso, T., Hanus, J., Evans, H. J.: Arch. Microbiol. *116*, 113 (1978)
78. Ruiz-Argüeso, T., Emerich, D. W., Evans, H. J.: Biochem. Biophys. Res. Com. *86*, 259 (1979)
79. Ruiz-Argüeso, T., Emerich, D. W., Evans, H. J.: Arch. Microbiol. *121*, 199 (1979)
80. Ruiz-Argüeso, T., Maier, R. J., Evans, H. J.: Appl. Environ. Microbiol. *37*, 582 (1979)
81. Carter, K. R., Jennings, N. T., Evans, H. J.: Can. J. Microbiol. *24*, 307 (1978)
82. Smith, L. A., Hill, S., Yates, M. G.: Nature *262*, 209 (1976)
83. Walker, C. C., Yates, G. M.: Biochimie *60*, 225 (1978)
84. Berlier, Y. M., Lespinat, P. A.: Arch. Microbiol. *125*, 67 (1980)
85. Lespinat, P. A., Berlier, Y. M.: FEMS Microbiol. Lett. *10*, 127 (1981)
86. Hungate, R. E.: The Rumen and its Microbes. New York, London: Academic Press 1966
87. Wolin, M. J.: Adv. Microb. Ecol. *3*, 49 (1979)
88. Müller, M.: Ann. Rev. Microbiol. *29*, 467 (1975)
89. Müller, M.: In: The Eucaryotic Microbial Cells. 30th Symp. Soc. Gen. Microbiol., p. 127, Cambridge: Univ. Press 1980
90. Yarlett, N., Williams, A. G., Lloyd, D.: In: Soc. General Microb. Quart. *8*, 130 (1981)
91. Zavarzin, G. A.: Vodorodnie bacterii i Karboksidobacterii. Moscow: Nauka 1978
92. Hanus, F. V., Maier, R. I., Evans, H. I.: Proc. Nat. Acad. Sci. USA *76*, 1788 (1979)
93. Malik, K. A., Schlegel, H. G.: FEMS Microbiol. Lett. *11*, 63 (1981)
94. Mah, R. A.: Ann. Rev. Microbiol. *31*, 309 (1977)
95. Balch, W. E. et al.: Microbiol. Rev. *43*, 260 (1979)
96. LeGall, J., Postgate, J. R.: Adv. Microb. Physiol. *10*, 81 (1973)
97. Badziong, W., Thauer, R. K., Zeikus, J. G.: Arch. Microbiol. *116*, 41 (1978)
98. Badziong, W., Ditter, B., Thauer, R. K.: ibid. *123*, 301 (1979)
99. Badziong, W., Thauer, R. K.: ibid. *125*, 167 (1980)
100. Widdel, F.: Anaerober Abbau von Fettsäuren und Benzoesäure durch neuisolierte Arten Sulfat-reduzierender Bakterien. Diss. Univ. Göttingen 1980
101. Balch, W. E. et al.: Int. J. Syst. Bacteriol. *27*, 355 (1977)
102. Adamse, A. D.: Antonie van Leeuwenhoek. J. Microbiol. Serol. *46*, 523 (1980)
103. Braun, K., Mayer, F., Gottschalk, G.: Arch. Microbiol. *128*, 288 (1981)
104. Macy, J., Kulla, H., Gottschalk, G.: J. Bacteriol. *125*, 423 (1976)
105. Bernhard, Th., Gottschalk, G.: Arch. Microbiol. *116*, 235 (1978)
106. Yamamoto, S., Shimoto, M.: J. Biochem. *84*, 673 (1978)
107. Chen, J.-S., Blanchard, D. K.: Biochem. Biophys. Res. Commun. *84*, 1144 (1978)
108. Chen, J.-S.: In: Hydrogenases: Their Catalytic Activity, Structure and Function (Schlegel, H. G., Schneider, K. eds.), p. 57. Göttingen: E. Goltze, 1978
109. Zacepin, S. S.: Mikrobiologija *49*, 489 (1980)
110. Zeikus, J. G.: Enzyme Microb. Technol. *1*, 243 (1979)
111. Morris, J. R.: Adv. Microbial. Physiol. *12*, 169 (1975)
112. Doelle, H. W.: Bacterial Physiology. New York, London: Academic Press 1969
113. Gottschalk, G.: Bacterial Metabolism. New York, Heidelberg, Berlin: Springer 1979
114. Bryant, M. P.: In: Microbiol. Energy Conversion (Schlegel, H., Barnea, J. eds.), p. 106. Göttingen: E. Goltze 1976
115. Walther, K., Hippe, H., Gottschalk, G.: Appl. Environ. Microbiol. *33*, 955 (1977)
116. McInerney, M. I., Bryant, M. P., Pfennig, N.: Arch. Microbiol. *122*, 129 (1979)
117. Bishop, N. I.: Ann. Rev. Plant Physiol. *17*, 185 (1966)
118. Lobanok, A. G., Babickaja, V. G.: Mikrobiologischeskij sintez belka na celluloze, Minsk: Nauka i technika 1976

119. Ohwaki, K., Hungante, R. E.: Appl. Environ. Microbiol. *33*, 1270 (1977)
120. Wolin, M. J., Miller, T. L.: Arch. Microbiol. *124*, 137 (1980)
121. Wolin, M. J.: In: Microbial Production and Utilization of Gases (Schlegel, H., Gottschalk, G., Pfennig, N. eds.), p. 141, Göttingen: E. Goltze 1976
122. Reddy, C. A. M., Bryant, M. P., Wolin, M. J.: J. Bacteriol. *110*, 126 (1972)
123. Reddy, C. A. M., Bryant, M. P., Wolin, M. J.: ibid. *110*, 133 (1972)
124. Glass, T. L., Bryant, M. P., Wolin, M. J.: ibid. *131*, 463 (1977)
125. Schaner, N. L., Ferry, J. G.: ibid. *142*, 800 (1980)
126. Zehnder, A. J. B., Huser, B. A., Brock, T. D.: Arch. Microbiol. *124*, 1 (1980)
127. Jungermann, K., Schön, G.: Arch. Microbiol. *99*, 109 (1974)
128. Douglas, M. U., Ward, F. B., Cole, J. A.: J. Gen. Microbiol. *80*, 557 (1974)
129. Chippaux, M., Pascal, M. C., Casse, F.: Eur. J. Biochem. *72*, 149 (1977)
130. Schön, G., Voelskow, H.: Arch. Microbiol. *107*, 87 (1976)
131. Dixon, R. O. D.: Biochimie *60*, 233 (1978)
132. Brill, W. J.: In: The Bacteria. Mechanisms of Adaptation (Gunsalus, I. C., Socatch, J. R., Ornston, L. N. eds.), Vol. VII, p. 85. New York: Acad. Press 1979
133. Brill, W. J.: Amer. Sci. *67*, 458 (1979)
134. Andersen, K., Shanmugam, K. T., Valentine, R. C.: In: Biological Solar Energy Conversion (Mitsui, A. et al. eds.), p. 339, New York, San Francisco: Academic Press 1977
135. Pinkwart, M. et al.: FEMS Microbiol. Lett. *6*, 177 (1979)
136. Karube, I. et al.: Biochim. Biophys. Acta *444*, 338 (1976)
137. Matsunaga, T., Karube, I., Suzuki, S.: Europ. J. App. Microbiol. Biotechn. *10*, 235 (1980)
138. Matsunaga, T., Karube, I., Suzuki, S.: Biotech. Bioeng. *22*, 2607 (1980)
139. Hatchikian, E. C., Chaigneau, M., LeGall, J.: In: Microbial Production and Utilization of Gases (Schlegel, H., Gottschalk, G., Pfennig, N. eds.), p. 109. Göttingen: E. Goltze 1976
140. Ben-Bassat, A., Lamed, R., Zeikus, J. G.: Arch. Microbiol. *146*, 192 (1981)
141. Chung, K. R.: Appl. Environ. Microbiol. *31*, 342 (1976)
142. Wiegel, J., Ljungdahl, L. G., Rawson, J. R.: J. Bacteriol. *39*, 800 (1979)
143. Iannotti, E. L. et al.: ibid. *114*, 1231 (1973)
144. Lamed, R., Zeikus, J. G.: ibid. *144*, 569 (1980)
145. Henderson, C.: J. Gen. Microbiol. *119*, 485 (1980)
146. Bryant, M. P. et al.: Appl. Environ. Microbiol. *33*, 1162 (1977)
147. Scherfringer, C. C., Linehan, B., Wolin, M. J.: Appl. Microbiol. *29*, 480 (1975)
148. Weimer, P. J., Zeikus, J. G.: Appl. Environ. Microbiol. *33*, 289 (1977)
149. Sineritz, F., Pirt, S. J.: J. Gen. Microbiol. *101*, 57 (1977)
150. Winter, J., Wolfe, R. S.: Arch. Microbiol. *121*, 97 (1979)
151. Vosjan, J. H.: Plant Soil. *43*, 317 (1975)
152. Schink, B., Schlegel, H. G.: Biochemie *60*, 297 (1978)
153. Vogels, G. D.: J. Microbiol. *45*, 347 (1979)
154. Belajev, S. S., Finkelstein, Z. I., Ivanov, M. V.: Mikrobiologija *44*, 309 (1975)
155. Brock, T.: Biol. of Microorganisms. Engelwood Cliffs, New Jersey: Prentis-Hall, Inc. 1979
156. Thauer, R.: In: Microbial Energy Conversion (Schlegel, H., Barnea, J. eds.), p. 201. Göttingen: E. Goltze 1976
157. Blanchard, G. C., Foley, R. T.: J. Electrochem. Soc. *118*, 1232 (1971)
158. Cenek, M.: Chem. Listy *62*, 927 (1968)
159. Bukin, V. I., Bichovski, V. J., Panzchava, E. S.: In: Vitamin B_{12} i ego primenenie v givotnovodstve. p. 9. Moskow: Nauka 1971
160. Burris, R. H.: In: Biol. Solar Energy Conversion (Mitsui, A. et al. eds.), p. 277. New York: Academic Press 1977
161. Renwick, G. M., Guimarro, C., Siegel, S. M.: Plant Physiol. *39*, 303 (1964)
162. Efimtsev, E. I., Boichenko, V. A., Litvin, F. F.: Dokl. Biophys. Akad. Nauk SSSR *220*, 986 (1975)
163. Pfennig, N., Trüper, H. G.: Bergey's Manual of Determinative Bacteriology. p. 24. Baltimore: Williams and Wilkins Comp. 1974
164. Trüper, H., Pfennig, N.: In: The Photosynthetic Bacteria (Clayton, R. K., Sistrom, W. R. eds.), p. 9. New York, London: Plenum Press 1978

165. Kondratieva, E. N.: In: Microbial Biochem. (Quayle, J. R. ed.), p. 117. Baltimore: Univ. Park Press 1979
166. Jones, O. T. G.: In: Microbial Energetics. 27th Symp. Soc. Gen. Microbiol., p. 151. Cambridge: Univ. Press 1977
167. Uffen, R. L.: In: The Photosynthetic Bacteria (Clayton, R. K., Sistrom, W. R. eds.), p. 857. New York, London: Plenum Press 1978
168. Madigan, M. T., Gest, H.: Arch. Microbiol. *117*, 119 (1978)
169. Kondratieva, E. N., Gorlenko, V. M.: Usp. Mikrobiol. *13*, 8 (1978)
170. Kampf, Ch., Pfennig, N.: Arch. Microbiol. *127*, 125 (1980)
171. Zurrer, H., Bachofen, R.: Experientia *36*, 1166 (1980)
172. Siefert, E., Pfennig, N.: Arch. Microbiol. *122*, 177 (1979)
173. Madigan, M. T., Gest, H.: J. Bacteriol. *137*, 524 (1979)
174. Yoch, D. C.: In: The Photosynthetic Bacteria (Clayton, R. K., Sistrom, W. R. eds.), p. 657. New York, London: Plenum Press 1978
175. Nordlund, S.: Studies on the nitrogenase system of the photosynthetic bacterium Rhodospirillum rubrum. Diss. Univ. Stockholm 1979
176. Kondratieva, E. N.: Photosynthetic Bacteria. London: Oldbourne Press 1965
177. Schön, G., Biedermann, M.: Biochim. Biophys. Acta *304*, 65 (1973)
178. Voelskow, H., Schön, G.: Arch. Microbiol. *119*, 129 (1978)
179. Voelskow, H., Schön, G.: ibid. *125*, 245 (1980)
180. Gorell, T. E., Uffen, R. L.: J. Bacteriol. *131*, 533 (1977)
181. Quadri, S. M. H., Hoare, D. S.: ibid. *95*, 2344 (1968)
182. Gogotov, I. N.: In: Abstracts Symp. on Pro. Photosynthetic Organisms (Drews, G. ed.), p. 118, Freiburg i. Br. 1973
183. Gogotov, I. N., Mitkina, T. V., Glinskij, V. P.: Mikrobiologija *43*, 586 (1974)
184. Uffen, R. L.: Proc. Nat. Acad. Sci. USA *73*, 3198 (1976)
185. Dashkevics, M. P., Uffen, R. L.: Intern. J. System. Bacteriol. *29*, 145 (1979)
186. Hageman, G.: Trends Biochem. Sci. *5*, 256 (1980)
187. Kondratieva, E. N., Gogotov, I. N., Gruzinskij, I. V.: Mikrobiologija *48*, 389 (1979)
188. Gogotov, I. N., Novikova, N. A.: ibid. *37*, 19 (1968)
189. Ensign, J. C.: In: Microbial Energy Conversion (Schlegel, H., Bornea, J. eds.), p. 455. Göttingen: E. Goltze 1976
190. Zürrer, H., Bachofen, R.: Appl. Environ. Microbiol. *37*, 789 (1979)
191. Schick, H.-J.: Arch. Mikrobiol. *75*, 89 (1971)
192. Schick, H.-J.: ibid. *75*, 110 (1971)
193. Schick, H.-J.: ibid. *75*, 102 (1971)
194. Hillmer, P., Gest, H.: J. Bacteriol. *129*, 724 (1977)
195. Hillmer, P., Gest, H.: ibid. *129*, 732 (1977)
196. Jouanneau, Y. et al.: ibid. *143*, 628 (1980)
197. Gogotov, I. N.: Dokl. Biol. Akad. Nauk SSSR *183*, 954 (1968)
198. Gogotov, I. N.: In: Enzyme Eng. Future Directions (Wingard, L. B., Berezin, I. V., Klyosov, A. A. eds.), p. 321. New York, London: Plenum Press 1980
199. Watanabe, K. et al.: Agric. Biol. Chem. *45*, 217 (1981)
200. Pashinger, H.: Arch. Microbiol. *101*, 379 (1974)
201. Kim, J. S., Ito, K., Takahashi, H.: Agric. Biol. Chem. *44*, 827 (1980)
202. Yoch, D. C.: J. Bacteriol. *140*, 987 (1979)
203. Song, H. et al.: Sci. Sinica *23*, 252 (1980)
204. Gogotov, I. N., Zorin, N. A.: Mikrobiologija *41*, 948 (1972)
205. Gogotov, I. N., Zorin, N. A., Bogorov, L. V.: ibid. *43*, 5 (1974)
206. Gogotov, I. N., Glinskij, V. P.: ibid. *42*, 983 (1973)
207. Zumft, W. F., Castillo, F.: Arch. Microbiol. *117*, 53 (1978)
208. Hillmer, P., Fahlbusch, K.: Arch. Microbiol. *122*, 213 (1979)
209. Sweet, W. J., Burris, R. H.: J. Bacteriol. *145*, 824 (1981)
210. Wall, J. D., Weaver, P. F., Gest, H.: Nature *258*, 630 (1975)
211. Weare, N. M.: Biochim. Biophys. Acta *502*, 486 (1978)
212. Wall, J. D., Gest, H.: J. Bacteriol. *137*, 1459 (1979)

213. Schestakov, S. et al.: In: 3rd Int. Symp. on Photosynthetic Prokaryotes. Abstracts., p. C8, Oxford 1979
214. Serebriakova, L. T., Teslia, E. A., Gogotov, I. N., Kondratieva, E. N.: Mikrobiologija *49*, 401 (1980)
215. Gogotov, I. N., Kondratieva, E. N.: Izvestija Akad. Nauk SSSR, seria biol. *1*, 161 (1969)
216. Olson, J. M., Thornber, J. P.: In: Membrane Proteins in Energy Transduction (Capaldi, R. A. ed.), p. 279. New York, Basel: Marcel Dekker 1979
217. Arnon, D. I., Yoch, D. C.: In: Biology of Nitrogen Fixation (Quiespel, A. ed.), p. 168. Amsterdam: North-Holland Publishing Co. 1974
218. Madigan, M. T., Wall, J. D., Gest, H.: Science *204*, 1429 (1979)
219. Siefert, E., Pfennig, N.: Arch. Microbiol. *125*, 73 (1980)
220. Macler, B. A., Pelroy, R. A., Bassham, J. A.: J. Bacteriol. *138*, 446 (1979)
221. Gest, H., Ormerod, I. G., Ormerod, K. S.: Arch. Biochem. Biophys. *97*, 21 (1962)
222. Weaver, P., Lien, S., Seibert, M.: Photobiol. Production of Hydrogen-a Solar Energy Conversion Option. Solar Energy Res. Inst. 1979
223. Knaff, D. B.: In: Photosynthetic Bacteria (Clayton, R. K., Sistrom, W. R. eds.), p. 629. New York, London: Plenum Press 1978
224. Pierson, B. K., Castenholz, R. W.: Arch. Microbiol. *100*, 5 (1974)
225. Palz, W., Chartier, P., Hall, D. O. (eds.): Energy from Biomass. London: Appl. Science Publ. LTD 1981
226. Madigan, M. T., Peterson, S. R., Brock, T. D.: Arch. Microbiol. *100*, 97 (1974)
227. Madigan, M. T., Brock, T. D.: FEMS Microbiol. Lett. *1*, 301 (1977)
228. Kondratieva, E. N., Gogotov, I. N.: Mikrobiologija *38*, 938 (1969)
229. Fogg, G. E. et al. (eds.): The Blue-green Algae. London, New York: Academic Press 1973
230. Stanier, R. J., Cohen-Bazire, G.: Ann. Rev. Microbiol. *31*, 225 (1977)
231. Rippka, R. et al.: J. Gen. Microbiol. *111*, 1 (1979)
232. Cohen, Y., Padan, E., Shilo, M.: J. Bacteriol. *123*, 855 (1975)
233. Padan, E.: Ann. Rev. Plant Phys. *30*, 27 (1979)
234. Padan, E.: Adv. Microbiol. Ecol. *3*, 1 (1979)
235. Benemann, J. R., Hallenbeck, P. C.: In: Hydrogenases: Their Catalytic Activity, Structure and Function (Schlegel, H., Schneider, K. eds.), p. 171. Göttingen: E. Goltze 1978
236. Peschek, G. A.: Arch. Microbiol. *119*, 313 (1978)
237. Peschek, G. A.: Biochim. Biophys. Acta *548*, 187 (1979)
238. Peschek, G. A.: FEBS Lett. *106*, 34 (1979)
239. Belkin, S., Padan, E.: Arch. Microbiol. *116*, 109 (1978)
240. Stewart, W. D. P. et al.: In: Proc. 4th Int. Congr. Photosynthesis (Hall, D. O., Coombs, J., Goodwin, T. W. eds.), p. 133. London: Biochem. Soc. 1978
241. Pankratova, E. M.: Izvestija Akad. Nauk SSSR, seria biol. *2*, 188 (1979)
242. Benemann, J. R., Weare, N. M.: Science *184*, 174 (1974)
243. Jones, L. W., Bishop, N. K.: Plant Physiol. *57*, 659 (1976)
244. Gogotov, I. N., Kosiak, A. V.: Mikrobiologija *45*, 586 (1976)
245. Gogotov, I. N., Kosiak, A. V., Krupenko, A. N.: ibid. *45*, 941 (1976)
246. Mitsui, A., Kumazawa, S.: In: Biol. Solar Energy Conversion (Mitsui, A. et al. eds.), p. 23. New York: Academic Press 1977
247. Tel-Or, E., Packer, L.: In: Hydrogenases: Their Catalytic Activity, Structure and Function (Schlegel, H., Schneider, K. eds.), p. 371. Göttingen: E. Goltze 1978
248. Benemann, J. R., Weare, N. M.: Arch. Microbiol. *101*, 401 (1974)
249. Benemann, J. R. et al.: Solar Energy Conversion Through Biophotolysis, p. 78. USB-SER and Report 1978
250. Kosiak, A. V.: In: 2nd Int. Symp. Growth of Microorganisms on C_1-compounds. Abstracts, p. 136. Pushchino: ONTI 1977
251. Asato, Y., Carr, N. G.: In: 3rd Int. Symp. Photosynthetic Prokaryotes. Abstracts, p. 10. Oxford 1977
252. Fogg, G. E.: Biologist *26*, 7 (1979)
253. Kerfin, W. et al.: In: Hydrogenases: Their Catalytic Activity, Structure and Function (Schlegel, H., Schneider, R. eds.), p. 381. Göttingen: E. Goltze 1978
254. Belkin, S., Padan, E.: FEBS Lett. *94*, 291 (1978)

255. Belkin, S., Padan, E.: In: Hydrogenases: Their Catalytic Activity, Structure and Function (Schlegel, H., Schneider, K. eds.), p. 387. Göttingen: E. Goltze 1978
256. Miyamoto, K., Hallenbeck, P. C., Benemann, J. R.:Apl. Environ. Microbiol. *37*, 3 (1979)
257. Miyamoto, K., Hallenbeck, P. C., Benemann, J. R.: ibid. *38*, 440 (1979)
258. Miyamoto, K., Hallenbeck, P. C., Benemann, J. R.: J. Ferment. Technol. *57*, 287 (1979)
259. Miyamoto, K., Miura, Y.: Process Biochem. *15*, 23 (1980)
260. Stranberg, G. W.: Eur. J. Appl. Microbiol. Biotech. *9*, 19 (1980)
261. Berchtold, M., Buchanan, R.: Arch. Microbiol. *123*, 227 (1979)
262. Newton, J. W.: Science *191*, 559 (1976)
263. Peters, G. A., Evans, W. R., Toria, R. E.: Plant Physiol. *59*, 119 (1976)
264. Peters, G. A., Toria, R. E., Lough, S. M.: ibid. *59*, 1021 (1977)
265. Bothe, H., Tennigkeit, J., Eisbrenner, G.: Arch. Microbiol. *114*, 43 (1977)
266. Bothe, H. et al.: Planta *133*, 237 (1977)
267. Laczkó, I. Z.: Z. Pflanzenphysiol. *100*, 241 (1980)
268. Lambert, G. R., Smith, G. D.: FEBS Lett. *83*, 159 (1977)
269. Lambert, G. R., Daday, A., Smith, G. D.: Appl. Environ. Microbiol. *38*, 530 (1979)
270. Lambert, G. R., Daday, A., Smith, G. D.: FEBS Lett. *101*, 125 (1979)
271. Jeffries, T. W., Timourian, H., Ward, R. L.: Appl. Environ. Microbiol. *35*, 704 (1978)
272. Hallenbeck, P. C., Kochian, L. V., Benemann, J. R.: Z. Naturforsch. *36*, 87 (1981)
273. Weissman, J. C., Benemann, J. R.: Appl. Environ. Microbiol. *33*, 123 (1977)
274. Lambert, G. R., Daday, A., Smith, G. D.: ibid. *38*, 521 (1979)
275. Schere, S., Kerfin, W., Böger, P.: J. Bacteriol. *141*, 1037 (1980)
276. Daday, F. R., Platz, A., Smith, G. D.: Appl. Environ. Microbiol. *34*, 478 (1977)
277. Lambert, G. R., Smith, G. D.: Arch. Biochem. Biophys. *205*, 36 (1980)
278. Peschek, G. A.: Arch. Microbiol. *123*, 81 (1979)
279. Peschek, G. A.: Biochem. Biophys. Acta *548*, 203 (1979)
280. Peschek, G. A.: Arch. Microbiol. *125*, 123 (1980)
281. Tel-Or, E., Luijk, L. W., Packer, L.: FEBS Lett. *78*, 49 (1977)
282. Tel-Or, E., Luijk, L. W., Packer, L.: Arch. Biochem. Biophys. *185*, 185 (1978)
283. Daday, A., Lambert, G. R., Smith, G. D.: Biochem. J. *177*, 139 (1979)
284. Nordlund, S.: Rapp. Ingenjörsvetenskapsakad *191*, 54 (1981)
285. Benemann, J. R.: In: Living Systems as Energy Converters (Buvet, R., Allen, M. J., Massue, J. P. eds.), p. 285. Amsterdam; Elsevier/North-Holland Biomed. Press 1977
286. Krampitz, L. O.: In: An Inquiry Into Biol. Energy Conversion (Hollaender, A. ed.), p. 22. Knoxville: Univ. Tennessee 1972
287. Ochiai, H. et al.: Proc. Nat. Acad. Sci. USA *77*, 2442 (1980)
288. Gaffron, H.: In: Plant Physiology (Steward, E. C. ed.), Vol. 1B, p. 185. New York, London: Academic Press 1960
289. Bishop, N. I.: Ann. Rev. Plant Physiol. *17*, 185 (1966)
290. Bishop, N. I., Frick, M., Jones, L. W.: Biol. Solar Energy Conversion (Mitsui, A. et al. eds.), p. 3. New York: Academic Press 1977
291. Ben-Amotz, A.: Plant Physiol. *56*, 72 (1975)
292. Hydrogen Obtained with the Help of Alga. Kagaku to Kogyo. Chem. and Chem. Snd *33*, 148 (1980)
293. Stewart, W. D. P. (ed.): Algal Physiology and Biochemistry. Oxford, London: Blackwell Sci. Publ. 1974
294. Persanov, V. M., Gogotov, I. N.: Mikrobiologija *47*, 212 (1978)
295. Persanov, V. M., Gogotov, I. N.: Fiziologia Rastenij *25*, 1139 (1978)
296. Yanushin, M. F.: ibid. *26*, 394 (1979)
297. Healey, F. P.: Planta *91*, 220 (1970)
298. Ochepkov, V. P., Krasnovsky, A. A.: Fisiologia Rastenij *19*, 1090 (1972)
299. Ochepkov, V. P., Krasnovsky, A. A.: ibid. *21*, 462 (1974)
300. Persanov, V. M., Gogotov, I. N.: ibid. *26*, 560 (1979)
301. Ochepkov, V. P. et al.: ibid. *25*, 821 (1978)
302. Krasnovsky, A. A.: Izvestija Acad. Nauk SSSR, seria biol. *5*, 650 (1977)
303. Krasnovsky, A. A.: In: Research in Photobiology (Castelani, I. ed.), p. 361. New York, London: Plenum Press 1977

304. Krasnovsky, A. A.: In: Photosynthesis in Relation to Model Systems (Barber, J. ed.), p. 282. Amsterdam: Elsevier/North-Holland Biomed. Press 1979
305. Stuart, T. S., Kaltwasser, H.: Planta *91*, 302 (1970)
306. Healey, F. P.: Plant Physiol. *45*, 153 (1970)
307. Kaltwasser, H., Stuart, T. S., Gaffron, H.: Planta *89*, 309 (1969)
308. McBride, A. C. et al.: In: Biol. Solar Energy Conversion. (Mitsui, A. et al. eds.), p. 77. New York: Academic Press 1977
309. Van Dijk, C., Veeger, C.: Europ. J. Biochem. *114*, 209 (1981)
310. Vinayakumar, M., Kessler, E.: Arch. Microbiol. *103*, 13 (1975)
311. Klein, V., Bentz, A.: Plant Physiol. *61*, 953 (1978)
312. Stuart, T. S., Gaffron, H.: Planta *106*, 91 (1972)
313. Stuart, T. S., Gaffron, H.: ibid. *106*, 101 (1972)
314. Efimcev, E. I. et al.: Dokl. Akad. Nauk SSSR, seria bioph. *226*, 45 (1976)
315. Krasna, A. I.: In: Methods in Enzymology Biomembranes (Fleisher, S., Packer, L. eds.), Vol. 3, p. 296. New York: Academic Press 1978
316. Greenbaum, E.: Biol. Solar Energy Conversion (Mitsui, A. et al. eds.), p. 101. New York: Academic Press 1977
317. Greenbaum, E.: Biotech. Bioeng. Symp. *10*, 1 (1980)
318. Pow, T., Krasna, A. I.: Arch. Biochem. Biophys. *194*, 413 (1979)
319. Schneider, K., Schlegel, H. G.: Biochim. Biophys. Acta *452*, 66 (1976)
320. Wang, R., Healey, F. P., Myers, J.: Plant Physiol. *48*, 108 (1971)
321. Kimura, K. et al.: Biochim. Biophys. Acta *567*, 96 (1972)
322. Peterson, R. B., Burris, R. H.: Arch. Microbiol. *116*, 125 (1978)
323. Mortenson, L. E., Chen, J.-S.: In: Microbial Production and Utilization of Gases (Schlegel, H., Gottschalk, G., Pfennig, N. eds.), p. 97. Göttingen: E. Goltze 1976
324. Graham, A.: Biochem. J. *197*, 283 (1981)
325. Matsunaga, T., Karube, I., Suzuki, S.: Biotech. Bioeng. *22*, 2607 (1980)
326. Zumft, W. G., Nordlund, S.: FEBS Lett. *127*, 79 (1981)
327. Gogotov, I. N., Zorin, N. A., Laurinavichene, T. V.: In: Hydrogenases: Their Catalytic Activity, Structure and Function (Schlegel, H., Schneider, K. eds.), p. 171. Göttingen: E. Goltze 1978
328. Schink, B., Schlegel, H. G.: Antonie van Leeuwenhoek. J. Microbiol. Serol. *46*, 1 (1980)
329. Hatchikian, E. C., Bruschi, M., Le Gall, J.: Biochem. Biophys. Res. Commun. *82*, 451 (1978)
330. Van der Westen, H. M., Mayhew, S. G., Veeger, G.: FEBS Lett. *86*, 122 (1978)
331. Glick, B. R., Martin, W. G., Martin, S. M.: Can. J. Microbiol. *26*, 1214 (1980)
332. Zorin, N. A., Gogotov, I. N.: Biochimia *45*, 1497 (1980)
333. Hiura, H. et al.: J. Biochem. *86*, 1151 (1979)
334. Van der Westen, H. M., Mayhew, S. G., Veeger, C.: FEMS Lett. *7*, 35 (1980)
335. Martin, S. M., Glick, B. R., Martin, W. G.: Can. J. Microbiol. *26*, 10 (1980)
336. Gogotov, I. N., Azova, L. G.: Mikrobiologija *45*, 28 (1976)
337. Mischustin, E. N., Emcev, V. T.: Pochvennie azotphiksiruushie bacterii roda Clostridium. Moscow: Nauka 1974
338. Veldkamp, H.: Antonie van Leeuwenhoek. J. Microbiol. Serol. *26*, 103 (1960)
339. Balaschova, V. V.: Mikoplasmi i gelezobacterii. Moskow: Nauka 1974
340. Walther, K., Hippe, H., Gottschalk, G.: Appl. Environ. Microbiol. *33*, 955 (1977)
341. Zehnder, A. J. B. et al.: Arch. Microbiol. *124*, 1 (1980)
342. Joklik, W. K.: Austral. J. Exptl. Biol. and Med. Sci. *28*, 321 (1950)
343. Andersen, K., Shanmugam, K.: J. Gen. Microbiol. *103*, 107 (1977)
344. Nakamura, H.: Acta Phytochim. *11*, 109 (1939)
345. Hoare, D. S., Hoare, S. L.: J. Bacteriol. *100*, 1124 (1969)
346. Pfennig, N., Trüper, H. G.: Phototrophic Bacteria. Neuherberg: GSF-Bericht M *32*, 143 (1969)
347. Fujita, Y., Myers, J.: Plant Physiol. *39*, 5 (1964)
348. Frenkel, A. W., Gaffron, H., Battley, E. H.: Nat. Biol. Bull. *99*, 157 (1950)
349. Llama, M. J. et al.: FEBS Lett. *98*, 342 (1979)
350. Lambert, G. R., Smith, G. D.: ibid. *83*, 159 (1977)
351. Gallon, J. R., Le Rue, W. G., Karz, W. G. W.: Can. J. Microbiol. *20*, 1633 (1974)

352. Kessler, E.: Arch. Mikrobiol. *61*, 77 (1968)
353. Stuart, T. S.: In: Workshop on Biosolar Conversion, p. 45. Bethesda: NSF/RANN Report 1973
354. Schubert, R. H. W.: Int. J. Syst. Bacteriol. *17*, 273 (1967)
355. Kessler, E.: In: Algal Physiology and Biochem. (Stewart, W. D. P. ed.), p. 456. Oxford: Blackwell 1974
356. Frenkel, A. W.: Arch. Biochem. Biophys. *38*, 219 (1952)
357. Gaffron, H., Rubin, Y.: J. Gen. Physiol. *26*, 219 (1942)
358. Gaffron, H.: Plant Phys. (Steward, E. C. ed.), Vol. 1B, p. 185. New York, London: Academic Press 1960
359. Kessler, E., Maifarth, H.: Arch. Mikrobiol. *37*, 215 (1960)
360. Frenkel, A. W., Rieger, C.: Nature *167*, 1030 (1951)
361. Yagi, T. et al.: J. Biochem. *79*, 661 (1976)
362. Yagi, T. et al.: J. Amer. Chem. Soc. *91*, 2801 (1969)
363. Schneider, K., Schlegel, H. G.: Arch. Microbiol. *112*, 229 (1977)
364. Bernhard, Th., Gottschalk, G.: In: Hydrogenases: Their Catalytic Activity, Structure and Function (Schlegel, H., Schneider, K. eds.), p. 199. Göttingen: E. Goltze 1978
365. Ackrell, B. A. C., Asato, R. H., Mower, H. F.: J. Bacteriol. *92*, 828 (1966)
366. Akagi, J. M., Campbell, L. L.: ibid. *82*, 927 (1961)
367. Buller, C. S., Akagi, J. M.: ibid. *88*, 440 (1964)
368. Vishniac, W., Trudinger, P. A.: Bacteriol. Rev. *26*, 168 (1962)
369. Mayhew, S. G., Dijk, C., van der Westen, M. M.: In: Hydrogenases: Their Catalytic Activity, Structure and Function (Schlegel, H., Schneider, K. eds.), p. 125. Göttingen: E. Goltze 1978
370. Fuchs, G. et al.: ibid. p. 83. Göttingen: E. Goltze 1978
371. Burns, R. C., Bulen, W. A.: Biochim. Biophys. Acta *105*, 437 (1965)
372. Schoenmaker, G. S., Oltmann, L. F., Stouthamer, A. H.: ibid. *567*, 511 (1979)
373. Sim, E., Colbeau, A., Vignais, P. M.: In: Hydrogenases: Their Catalytic Activity, Structure and Function (Schlegel, H., Schneider, K. eds.), p. 269. Göttingen: E. Goltze 1978
374. Gogotov, I. N., Netrusov, A. I., Kondratieva, E. N.: Mikrobiologija *44*, 779 (1975)
375. Emnova, E. E., Romanova, A. K.: ibid. *46*, 619 (1977)
376. Atkinson, D. E., McFadden, B. A.: J. Biol. Chem. *210*, 885 (1954)
377. Adams, M. W. W., Hall, D. O.: Biochem. Biophys. Res. Commun. *77*, 730 (1977)
378. Gillum, W. O. et al.: P. Amer. Chem. Soc. *99*, 584 (1977)
379. Kerfin, W. et al.: In: Hydrogenases: Their Catalytic Activity, Structure and Function (Schlegel, H., Schneider, K. eds.), p. 381. Göttingen: E. Goltze 1978
380. Gogotov, I. N. et al.: Biochim. Biophys. Acta *523*, 385 (1978)
381. Kakuno, T., Kaplan, N. O., Kamen, M. D.: Proc. Nat. Acad. Sci. USA *74*, 861 (1977)
382. Gitlitz, P. H., Krasna, A. J.: Biochem. *14*, 2561 (1975)
383. Colbeau, A. et al.: 3rd Int. Symp. on Photosynthetic Prokaryotes, p. B3. Oxford 1979
384. Weiss, A. R., Schlegel, H. G.: FEMS Microb. Lett. *8*, 173 (1980)
385. Hashke, R. H., Campbell, L. L.: J. Bacteriol. *105*, 249 (1971)
386. Okura, I., Nakamura, K., Keii, T.: J. Mol. Catal. *4*, 453 (1978)
387. Bell, G. R. et al.: Biochimie *60*, 315 (1978)
388. Van Dijk, C. et al.: Eur. J. Biochem. *107*, 251 (1980)
389. Sim, E., Vignais, P. M.: Biochim. Biophys. Acta *570*, 43 (1979)
390. Schneider, K. et al.: ibid. *578*, 445 (1979)
391. Schink, B., Schlegel, H. G.: ibid. *567*, 315 (1979)
392. Adams, M. W. W., Hall, D. O.: Arch. Biochem. Biophys. *195*, 730 (1979)
393. Van Heerikhuizen, H. et al.: In: Hydrogenases: Their Catalytic Activity, Structure and Function (Schlegel, H., Schneider, K. eds.), p. 151. Göttingen: E. Goltze 1978
394. Erbes, D. L., Burris, R. H.: Biochim. Biophys. Acta *525*, 45 (1978)
395. Le Gall, J. et al.: ibid. *234*, 525 (1971)
396. Gogotov, I. N., Zorin, N. A., Kondratieva, E. N.: Biokhimija *41*, 836 (1976)
397. Yagi, T.: J. Biochem. *68*, 649 (1970)
398. Yagi, T., Honya, M., Tamiya, N.: Biochim. Biophys. Acta *153*, 699 (1968)
399. Sadana, J. C., Jagannathan, A. V.: ibid. *19*, 440 (1956)

400. Yagi, T. et al.: J. Biochem. *78*, 443 (1975)
401. Schengrund, C., Krasna, A. I.: Biochim. Biophys. Acta *185*, 322 (1969)
402. Pfitzner, J., Linke, H. A. B., Schlegel, H. G.: Arch. Microbiol. *71*, 67 (1970)
403. Gruzinsky, I. V., Gogotov, I. N., Bechina, E. M., Semenov, J. V.: Mikrobiologija *46*, 625 (1977)
404. Pinchukova, E., Varfolomeev, S. D., Kondratieva, E. N.: Biokhimia *44*, 605 (1979)
405. Arp., D. J., Burris, R. H.: Biochim. Biophys. Acta *570*, 221 (1979)
406. Winter, H. C., Burris, R. H.: Ann. Rev. Biochem. *45*, 42 (1976)
407. Shah, V. K., Brill, W. J.: Biochim. Biophys. Acta *305*, 445 (1973)
408. Eady, R. R.: In: The Evolution of Metalloenzymes, Metalloproteins and Related Materials (Leigh, G. ed.), p. 67. Brighton: Univ. Sussex 1977
409. Emerich, D. W., Burris, R. H.: Biochim. Biophys. Acta *536*, 182 (1978)
410. Eady, R. R. et al.: Biochem. J. *128*, 655 (1972)
411. Berndt, H., Lowe, D. J., Yates, M. G.: Eur. J. Biochem. *86*, 133 (1978)
412. Israel, D. W. et al.: J. Biol. Chem. *249*, 500 (1974)
413. Aseeva, K. B. et al.: Izvestija Akad. Nauk SSSR seria biol. *5*, 751 (1973)
414. Nordlund, S., Eriksson, U., Baltscheffsky, H.: Biochim. Biophys. Acta *504*, 248 (1978)
415. Evans, M. C. W., Tefler, A., Smith, R. V.: ibid. *310*, 344 (1973)
416. Hallenbeck, P. C., Kostel, P. I., Benemann, I. R.: Eur. J. Biochem. *98*, 275 (1979)
417. Ludden, P. W., Burris, R. H.: Biochem. J. *175*, 251 (1978)
418. Benemann, J. R. et al.: Proc. Nat. Acad. Sci. USA *70*, 2317 (1973)
419. Krasnovsky, A. A. et al.: Dokl. Akad. Nauk SSSR, seria biophys. *225*, 711 (1975)
420. Rao, K. K., Rosa L., Hall, D. O.: Biochem. Biophys. Res. Com. *68*, 21 (1976)
421. Fry, I. et al.: Z. Naturforsch. *32*, 110 (1977)
422. Persanov, V. M. et al.: Fisiologia Rastenij *24*, 699 (1977)
423. Hoffmann, D., Thauer, R., Trebst, A.: Z. Naturforsch. *32*, 257 (1977)
424. Hall, D. O. et al.: In: Int. Compendium Alternative Energy Sources (Veziroglu, T. ed.). Vol. 8, p. 3675. Washington: Hemisphere Publ. Corp. 1978
425. Rao, K. K., Gogotov, I. N., Hall, D. O.: Biochimie *60*, 291 (1978)
426. Krasnovsky, A. A. et al.: Mol. Biol. *14*, 287 (1980)
427. Fitzgerald, M. P. et al.: Biochem. J. *192*, 665 (1980)
428. Serebriakova, L. T., Zorin, N. A., Gogotov, I. N.: Biokhimija *42*, 740 (1977)
429. Chen, J.-S., Mortenson, L. E.: Biochim. Biophys. Acta *37*, 283 (1974)
430. Yagi, T., Endo, A., Tsuji, K.: In: Hydrogenases: Their Catalytic Activity, Structure and Function (Schlegel, H., Schneider, K. eds.), p. 107. Göttingen: E. Goltze 1978
431. Llama, M. J. et al.: Eur. J. Biochem. *144*, 89 (1981)
432. Erbes, D. L., Burris, R. H., Orme-Johnson, W. H.: Proc. Nat. Acad. Sci. USA *72*, 4795 (1975)
433. Que, J., Holm, R. H., Mortenson, L. E.: J. Amer. Chem. Soc. *97*, 163 (1975)
434. Schneider, K., Cammack, R.: In: Hydrogenases: Their Catalytic Activity, Structure and Function (Schlegel, H., Schneider, K. eds.), p. 221. Göttingen: E. Goltze 1978
435. Holm, R. H.: Endeavour *34*, 38 (1975)
436. Van der Werf, A., Yates, M. G.: In: Hydrogenases: Their Catalytic Activity, Structure and Function (Schlegel, H., Schneider, K. eds.), p. 307. Göttingen: E. Goltze 1978
437. Kimura, K. et al.: Biochim. Biophys. Acta *567*, 96 (1979)
438. Krasna, A. I., Rittenberg, D.: J. Amer. Chem. Soc. *76*, 3015 (1954)
439. Bone, D. H.: Biochim. Biophys. Acta *67*, 589 (1963)
440. Varfolomeev, S. D. et al.: Mol. Biol. *12*, 63 (1978)
441. Gillum, W. O. et al.: Amer. Chem. Soc. *99*, 584 (1977)
442. Mortenson, L. E.: Biochimie *60*, 219 (1978)
443. Nakos, I., Mortenson, L. E.: Biochim. Biophys. Acta *227*, 576 (1971)
444. Berezin, I. V. et al.: Dokl. Akad. Nauk SSSR, seria biol. *220*, 237 (1975)
445. Toaj, Ch. D. et al.: Mol. Biol. *10*, 452 (1976)
446. Henry, L. E. A. et al.: FEBS Lett. *122*, 211 (1980)
447. Schneider, K., Schlegel, H. G.: Biochem. J. *193*, 99 (1981)
448. Gogotov, I. N., Kulakova, S. M.: Usp. Microbiol. *16*, 30 (1981)
449. Yagi, T.: J. Biochem. *68*, 649 (1970)

450. Erbes, D. L., King, D. Gibbs, M.: Plant Physiol. *63*, 1138 (1979)
451. McBride, A. C. et al.: In: Biol. Solar Energy Conversion (Mitsui, S. et al. eds.), p. 77. New York: Academic Press 1977
452. Weiss, A. R., Schneider, K., Schlegel, H. G.: Curr. Microbiol. *3*, 317 (1980)
453. Pinchukova, E. E., Varfolomeev, S. D., Berezin, I. V.: Dokl. Akad. Nauk SSSR, seria biol. *236*, 1253 (1977)
454. Schink, B., Probst, I.: Biochem. Biophys. Res. Commun. *95*, 1563 (1980)
455. Zorin, N. A., Gogotov, I. N., Kondratieva, E. N.: FEMS Microbiol. Lett. *5*, 301 (1979)
456. Karube, I. et al.: Biochim. Biophys. Acta *444*, 338 (1976)
457. Lappi, D. A. et al.: Biochem. Biophys. Res. Commun. *69*, 878 (1976)
458. Yagi, T.: In: Biol. Solar Energy Conversion (Mitsui, A. et al. eds.), p. 61. New York: Academic Press 1977
459. Simon, H., Egeror, P., Grinther, H.: In: Hydrogenases: Their Catalytic Activity, Structure and Function (Schlegel, H., Schneider, K. eds.), p. 235. Göttingen: E. Goltze 1978
460. Varfolomeev, S. D., Bachurin, S. D., Toaj, Ch. D.: Mol. Biol. *11*, 423 (1977)
461. Berenson, J. A., Benemann, J. R.: FEBS Lett. *76*, 105 (1977)
462. Klibanov, A. M., Kaplan, N. O., Kamen, M. D.: Proc. Nat. Acad. Sci. USA *75*, 3640 (1978)
463. Tabillion, R., Weber, F., Kaltwasser, H.: Arch. Microbiol. *124*, 131 (1980)
464. Gogotov, I. N.: Biochimie *60*, 267 (1978)
465. Sim, E., Vignais, P. M.: ibid. *60*, 307 (1978)
466. Schlegel, H. G., Eberhardt, U.: Adv. Microb. Physiol. *7*, 205 (1972)
467. Aggag, M., Schlegel, H. G.: Arch. Mikrobiol. *88*, 209 (1973)
468. Maier, E. J., Hanus, D. J., Evans, H. J.: J. Bacteriol. *137*, 824 (1979)
469. Simpson, F. B., Maier, R. J., Evans, H. J.: Arch. Microbiol. *123*, 1 (1979)
470. Partridge, C. D. P. et al.: J. Gen. Microbiol. *119*, 313 (1980)
471. Packer, L.: In: Methods in Enzymology. Photosynthesis and Nitrogen Fixation. *69*, 625 (1980)
472. Colbeau, A., Chabert, J., Vignais, P. M.: In: Hydrogenases: Their Catalytic Activity, Structure and Function (Schlegel, H., Schneider, K. eds.), p. 183. Göttingen: E. Goltze 1978
473. Aggag, M., Schlegel, H.: Arch. Microbiol. *100*, 25 (1974)
474. Aragno, M., Schlegel, H. G.: J. Syst. Bacteriol. *28*, 112 (1978)
475. Aragno, M., Schlegel, H. G.: Arch. Microbiol. *116*, 221 (1978)
476. Berndt, H., Wölfle, D.: In: Hydrogenases: Their Catalytic Activity, Structure and Function (Schlegel, H., Schneider, K. eds.), p. 327. Göttingen: E. Goltze 1978
477. Sweet, W. J., Burris, R. H.: J. Bacteriol. *145*, 824 (1981)
478. Evans, H. J. et al.: In: Hydrogenases: Their Catalytic Activity, Structure and Function (Schlegel, H., Schneider, K. eds.), p. 287. Göttingen: E. Goltze 1978
479. Burris, R. H.: In: Biol. Solar Energy Conversion (Mitsui, A. et al. eds.), p. 275. New York: Academic Press 1977
480. Newton, W., Postgate, J. R., Rodriguez-Barrueco, C.: Recent Development in Nitrogen Fixation, p. 622. London: Academic Press 1977
481. Lichtenschtein, G. I.: Mnogojadernie okislitelno-vosstanovitelnie metallofermenti. Moscow: Nauka 1979
482. Bulen, W. A., Burns, R. C., Le Comte, J. R.: Proc. Nat. Acad. Sci. USA *53*, 532 (1965)
483. Zumft, W. G.: In: Structure and Bonding. (Dunitz, J. D. et al. eds.), Vol. 29, p. 1. Berlin: Springer 1976
484. Mortenson, L. E.: In: Current Topics in Cellular Regulation (Horecker, B. L., Stadtman, E. R. eds.), Vol. 13, p. 179. New York, London: Academic Press 1978
485. Zumft, W. G., Mortenson, L. E.: Biochim. Biophys. Acta *416*, 1 (1975)
486. Hageman, R. V., Burris, R. H.: Proc. Nat. Acad. Sci. USA *75*, 2699 (1978)
487. Ludden, P. W., Burris, R. H.: ibid. *76*, 6201 (1979)
488. Dalton, H.: In: Microbial Biochem. (Quayle, G. R. ed.), p. 227. Baltimore: Univ. Park Press 1979
489. Ljones, T.: FEBS Lett. *98*, 1 (1979)
490. Shah, V. K., Chisnell, J. R., Brill, W. J.: Biochem. Biophys. Res. Commun. *81*, 232 (1978)
491. Benemann, J. R., Valentine, R. C.: Adv. Microb. Physiol. *8*, 59 (1972)

492. Ludden, P. W., Burris, R. H.: Science *194*, 424 (1976)
493. Nordlund, S., Eriksson, U., Baltscheffsky, H.: Biochim. Biophys. Acta *462*, 187 (1977)
494. Yoch, D. C., Cantu, M.: J. Bacteriol. *142*, 899 (1980)
495. Rivera-Ortiz, J. M., Burris, R. H.: ibid. *123*, 537 (1975)
496. Thorneley, R. N. F., Cornish-Bowden, A.: Biochem. J. *165*, 255 (1977)
497. Schilov, A. E.: Usp. chimii *43*, 863 (1974)
498. Ahmad, M. H.: J. Gen. and Appl. Microbiol. *24*, 271 (1978)
499. Meyer, J., Kelley, B. C., Vignais, P.: FEBS Lett. *85*, 224 (1978)
500. Meyer, J., Vignais, P. M.: Biochem. Biophys. Res. Commun. *89*, 353 (1979)
501. Jones, B. L., Monty, K. I.: J. Bacteriol. *139*, 1007 (1979)
502. Stacey, G. et al.: ibid. *137*, 321 (1979)
503. Stewart, W. D. P.: Endeavour New Ser. *2*, 170 (1978)
504. Nielson, A. H., Nordlund, S.: J. Gen. Microbiol. *91*, 53 (1975)
505. Nordlund, S., Eriksson, U.: Biochim. Biophys. Acta *547*, 429 (1979)
506. Kosiak, A. V., Gogotov, I. N., Kulakova, S. M.: Mikrobiologija *47*, 605 (1978)
507. Yoch, D. C.: Biochem. J. *187*, 273 (1980)
508. Nordlund, S., Eklund, R.: In: 3rd Int. Symp. on Photosynthetic Prokaryotes. Abstracts, p. B47. Oxford 1979
509. Kelley, B. C., Khanna, S., Nicholas, D. J.: Arch. Microbiol. *127*, 77 (1980)
510. Colbeau, A., Kelley, B. C., Vignais, P. M.: J. Bacteriol. *144*, 141 (1980)
511. Dixon, R. O. D.: Nature *262*, 173 (1976)
512. Emerich, D. W. et al.: J. Bacteriol. *137*, 153 (1979)
513. Albrecht, S. L. et al.: Science *203*, 1255 (1979)
514. Partridge, C. D. P. et al.: J. Gen. Microbiol. *119*, 313 (1980)
515. Peterson, R. B., Burris, R. H.: Arch. Microbiol. *116*, 125 (1978)
516. Siefert, E., Pfennig, N.: Biochemie *60*, 261 (1978)
517. Simon, H., Egeror, P., Grinther, H.: In: Hydrogenases: Their Catalytic Activity, Structure and Function (Schlegel, H., Schneider, K. eds.), p. 235. Göttingen: E. Goltze 1978
518. Berezin, I. V., Varfolomeev, S. D.: In: Appl. Biochem. Bioeng (Wingard, L. B. et al. eds.), Vol. 2, Enzyme Techn., p. 259. New York, San Francisco, London: Academic Press 1979
519. Rao, K. K., Hall, D. O.: In: Photosynthesis in Relation to Model Systems (Barber, J. ed.), p. 299. Amsterdam: Elsevier/North-Holland Biomed. Press 1979
520. Zamaraev, K. I., Parmon, V. N.: Catal. Rev., Sci. Eng. *22*, 261 (1980)
521. McInerney, M. J. et al.: Appl. Environ. Microbiool. *41*, 1029 (1981)
522. Sachno, O. N., Ivanovskij, R. N., Kondratieva, E. N.: Mikrobiologija *50*, 607 (1981)
523. Kondratieva, E. N., Ivanovsky, R. N., Krasilnikova, E. N.: In: Soviet Scient. Rev., p. 325. New York: Cordon and Breach 1981
524. Zumft, W. G., Arp, D. J.: Naturwissenschaften *68*, 424 (1981)
525. Houchins, J. P., Burris, R. H.: J. Bacteriol. *146*, 215 (1981)
526. Belkin, S., Rao, K. K., Hall, D. O.: Biochem. Int. *3*, 301 (1981)
527. Van Heerikhuizen, H. et al.: Biochim. Biophys. Acta *657*, 26 (1981)
528. Serebryakova, L. T., Gogotov, I. N.: Prikl. Biochem. Mikrob. *27*, 555 (1981)
529. Colbeau, A., Vignais, P.: Biochim. Biophys. Acta *662*, 271 (1981)
530. Gahalin, S. S.: In: Metals and Miner. Rev. *19*, 8 (1980)
531. Tomas, J., Tuli, R.: Indian J. Microbiol. *20*, 259 (1980)
532. Walker, C. C., Partridge, C. D., Yates, M. G.: J. Gen. Microb. *124*, 317 (1981)
533. Ludden, P. W., Burris, R. H.: Arch. Microbiol. *130*, 155 (1981)
534. Hochman, A., Burris, R. H.: J. Bacteriol. *147*, 492 (1981)
535. Alef, K., Arp, D. I., Zumft, W. G.: Arch. Microbiol. *130*, 138 (1981)
536. Salminen, S. O.: Biochim. Biophys. Acta 658, *1* (1981)
537. Ruiz-Argüeso, T. et al.: Arch. Microbiol. *128*, 285 (1981)
538. Volpon, G. T. et al.: ibid. *128*, 371 (1981)
539. Fridrich, B. et al.: J. Bacteriol. *145*, 1144 (1981)
540. Baresi, L., Wolfe, R. S.: Appl. Environ. Microbiol. *41*, 388 (1981)
541. Bowien, B., Schlegel, H. G.: Ann. Rev. Microbiol. *35*, 405 (1981)
542. Kayano, H. et al.: Biochim. Biophys. Acta *638*, 80 (1981)

Author Index Volumes 1–28